Fertility Preservation in Males

Emre Seli • Ashok Agarwal
Editors

Fertility Preservation in Males

Emerging Technologies and Clinical Applications

 Springer

Editors
Emre Seli, MD
Department of Obstetrics,
  Gynecology, and Reproductive Sciences
Yale University School of Medicine
New Haven, CT, USA

Ashok Agarwal, PhD, HCLD (ABB)
Center for Reproductive Medicine
Glickman Urological and Kidney Institute
OB-GYN and Women's Health Institute
Cleveland Clinic, Cleveland, OH, USA

ISBN 978-1-4614-5619-3
DOI 10.1007/978-1-4614-5620-9
Springer New York Heidelberg Dordrecht London

Library of Congress Control Number: 2012948986

Printed on acid-free paper

Springer is part of Springer Science+Business Media (www.springer.com)

Contents

Protocols

Contributors

Ashok Agarwal Center for Reproductive Medicine, Glickman Urological and Kidney Institute, OB-GYN and Women's Health Institute, Cleveland Clinic, Cleveland, OH, USA

Desiderio Avila Jr. Scott Department of Urology, Baylor College of Medicine, Houston, TX, USA

Ciler Celik-Ozenci Department of Histology and Embryology, School of Medicine, Akdeniz University, Campus, Antalya, Turkey

Fnu Deepinder Department of Endocrinology, Diabetes and Metabolism, Cedars Sinai Medical Center, Los Angeles, CA, USA

Ina Dobrinski Department of Comparative Biology and Experimental Medicine, University of Calgary, Calgary, AB, Canada

Bhushan K. Gangrade IVF Laboratory, Center for Reproductive Medicine, Orlando, FL, USA

Paul Gittens Scott Department of Urology, Baylor College of Medicine, Houston, TX, USA

Kathleen Hwang Scott Department of Urology, Baylor College of Medicine, Houston, TX, USA

Larry I. Lipshultz Lester and Sue Smith Chair in Reproductive Medicine, Baylor College of Medicine, Houston, TX, USA

Division of Male Reproductive Medicine and Surgery, Scott Department of Urology, Baylor College of Medicine, Houston, TX, USA

Ana Isabel Marques Mari Stem Cell Bank, Prince Felipe Research Centre, Valencia, Spain

William Murk Department of Obstetrics, Gynecology, and Reproductive Sciences, Yale University School of Medicine, New Haven, CT, USA

Queenie V. Neri The Ronald O. Perelman and Claudia Cohen Center for Reproductive Medicine, Weill Cornell Medical College, New York, NY, USA

Carolina Ortega Centre for Reproductive Medicine, University Hospital, Dutch-speaking Brussels Free University, Brussels, Belgium

Gianpiero D. Palermo The Ronald O. Perelman and Claudia Cohen Center for Reproductive Medicine, Weill Cornell Medical College, New York, NY, USA

Pasquale Patrizio Yale Fertility Center and REI Medical Practice, New Haven, CT, USA

Marcia Riboldi Stem Cell Bank, Prince Felipe Research Centre, Valencia, Spain

Jose R. Rodriguez-Sosa Department of Comparative Biology and Experimental Medicine, University of Calgary, Calgary, Canada

Zev Rosenwaks Reproductive Medicine in Obstetrics and Gynecology, Weill Cornell Medical College, New York, NY, USA

The Ronald O. Perelman and Claudia Cohen Center for Reproductive Medicine, Weill Cornell Medical College, New York, NY, USA

Stefan Schlatt Center of Reproductive Medicine and Andrology, University Hospital of Munster, Münster, Germany

Emre Seli Department of Obstetrics, Gynecology, and Reproductive Sciences, Yale University School of Medicine, New Haven, CT, USA

Carlos Simón Stem Cell Bank, Pricne Felipe Research Centre, Valencia, Spain

Pankaj Talwar ART Centre, Army Hospital Research and Referral, Dhaula Kuan, New Delhi, India

Herman Tournaye Centre for Reproductive Medicine, University Hospital, Dutch-Speaking Brussels Free University, Brussels, Belgium

Chapter 1
The Epidemiology of Fertility Preservation

William Murk and Emre Seli

In 2007, it was estimated that there were 11.2 million cancer survivors living in the USA, of whom 450,000 were of reproductive age [1]. Although cancer is commonly thought to be disease of the elderly, 9.5% of all cancers are diagnosed before the age of 45 in the USA [2]. Due to surgical removal of reproductive organs or gonadotoxic effects from cancer treatment, survivors will likely face compromised fertility. For those who do not have surgical sterility, it has been estimated that survivors of childhood cancer have an overall reduction of 46% in likelihood of ever siring a pregnancy (among men) and 19% in likelihood of ever being pregnant (among women) [3, 4]. The prospect of partial or total infertility can significantly add to anxiety and emotional strain during disease management, and may also compromise long-term quality of life [5]. To offset these risks, patients can be offered several options for fertility preservation, including conservative cancer management, and cryopreservation of sperm, embryos, or ovarian and testicular tissue.

In this chapter, we describe the epidemiology of factors relevant to fertility preservation. We begin by providing a description of demographic trends that illustrates the increasing need for fertility preservation, including trends in cancer survivorship and age of first pregnancy. This is followed by a brief overview of risk factors for infertility after cancer therapy. A summary of known health risks to pregnant cancer survivors and their offspring will be presented. Finally, we consider the current state of awareness and attitudes of fertility preservation among patients and medical providers.

E. Seli, M.D. (✉)
Department of Obstetrics, Gynecology, and Reproductive Sciences,
Yale University School of Medicine, New Haven, CT, USA
e-mail: emre.seli@yale.edu

E. Seli and A. Agarwal (eds.), *Fertility Preservation in Males: Emerging Technologies and Clinical Applications*, DOI 10.1007/978-1-4614-5620-9_1,
© Springer Science+Business Media New York 2012

Demographics of Fertility Preservation

Cancer Incidence

Cancer is not an uncommon disease among reproductive men and women. In the USA, the probability of developing any cancer by the age of 40 was estimated in 2004–2006 to be 2.36% in women, and 1.51% in men [6]. The age-adjusted incidence rates (AAIRs; see Table 1.1 for a definition) of this age group in 2007 were 40.7 malignant cases per 100,000 among men, and 59.8 per 100,000 among women [7]. At the age of 49, these numbers rise to 75.7 and 121.6 cases per 100,000, respectively [7]. The top ten sites of cancer by incidence for women aged 0–40 and men aged 0–49 are shown in Fig. 1.1a, b, respectively. Reproductive tissues ranked among these top ten sites, including the cervix uteri, corpus uteri, and ovary for women; and the testis and prostate for men. Among men, the male genital system as a whole had the highest cancer incidence rate of any general tissue system (AAIR, 13.6 cases per 100,000), while genital system ranked third among women (AAIR, 7.6 cases per 100,000), within these age groups [7]. Overall, the incidence rate of cancer by any site among ages less than 65 is declining, with an age-adjusted annual percent change (APC; see Table 1.1 for definition) of −0.6 for the period 1998–2007 [2]. By site, the incidence of cancer in reproductive tissues is declining for some sites, including the cervix uteri (APC, −3.2), ovary (APC, −1.6), and prostate (APC, −0.1; not significantly different from 0) [2]. However, the incidence among other reproductive tissues is increasing, including the corpus uteri (APC, 0.4) and testis (APC, 1.0).

Childhood and adolescent cancers (i.e., those diagnosed before the age of 20) are relatively rare, with AAIRs of 17.2 malignant cases per 100,000 among males, and 16.2 per 100,000 among females [7]. Nevertheless, these cancers are a significant public health problem. As of 1 January 2007, it was estimated that there were 269,403 people living in the USA who were diagnosed with cancer before the age of 20 (not including those diagnosed more than 31 years ago, due to incomplete data before 1975) [7]. Of these, 209,957 have been alive for at least 5 years since their diagnosis. Between 2003 and 2007, there were a total of 19,257 new cases of cancer for patients below the age of 20, of which 12,424 were diagnosed at a prepubertal age (0–14) [2]. The top ten sites of cancer for girls and boys aged 0–14 are shown in Fig. 1.2a, b, respectively. Cancers of the genital system are quite rare before the age of 20, with AAIRs of 1.4 cases per 100,000 among males, and 0.5 cases per 100,000 among females [7]. Before the age of 15, the AAIRs of these cancers are 0.4 per 100,000 among boys, and effectively zero among girls [7]. Notably, while the incidence rates of most adult cancers are decreasing, the incidence rate of childhood/adolescent cancers is increasing (APC between 1975 and 2007, 0.6) [2]. AAIRs have climbed from 12.9 cases per 100,000 in 1975, to 15.7 per 100,000 in 1995, and to 16.7 per 100,000 in 2007 [2].

Table 1.1 Definitions of commonly used terms in epidemiology

Term	Abbreviation	Definition	Examples and interpretations
Relative risk, odds ratio, hazard ratio, or incidence rate ratio	RR, OR, HR, or IRR	These are measures of the strength of an association between an exposure and an outcome. Specifically, the ratio of the risk, odds, hazard rate, or incidence rate of an outcome in an exposed group to that in an unexposed group	E.g., RR=2.5: an exposed individual has a 2.5 times greater *risk* of outcome, when compared with an unexposed individual RR>1: exposure increases risk of outcome RR=1: exposure does not affect risk RR<1: exposure reduces risk
95% confidence interval	95% CI	A measure of the error (uncertainty) of some statistically distributed parameter (e.g., for RR, OR, HR, IRR, prevalence, etc.). If a parameter were to be independently measured 100 times in a random sample, the parameter estimate would fall within the calculated confidence interval 95 out of those 100 times	E.g., RR=2.5 (95% CI: 1.9–3.1). The RR is measured to be 2.5, but if repeatedly measured, it would fall within the range 1.9–3.1 95% of the time. Less formally, there is 95% certainty that the true measure of association is within the range 1.9–3.1. If the 95% CI does not include the null (1.0), then it can be concluded that the effect is significantly different than expected from chance, with a false-positive probability of 5%
Incidence rate	IR	Number of new cases per population size in a given time period	E.g., 27 cases per 100,000 people in 2007
Age-adjusted incidence rate AAIR	AAIR	Incidence rate that a population would have if it had the same age structure as some reference population (often, the US census population is used as the referent for American studies). Allows comparisons of incidence rates between populations that have different age distributions, if they are adjusted using the same reference population. Reduces the potential for confounding by age	E.g., Population 1 (mostly elderly): crude (unadjusted) IR of 78 cases per 100,000. Population 2 (mostly youth): crude (unadjusted) IR of 23 cases per 10,000. From crude analysis, it appears that population 1 has a greater risk. However, this may be due to an effect of age (e.g., elderly are at higher risk) rather than population (e.g., population 1 is at higher risk). If this is true, then using an age-adjusted incidence rate will reveal no or reduced difference between the populations
Annual percent change	APC	An estimate of percent change in a rate over time, assuming that rates change at a constant percentage rate	APC: −8.6 for 1999–2003: there was an annual 8.6% reduction in rate between 1999 and 2003

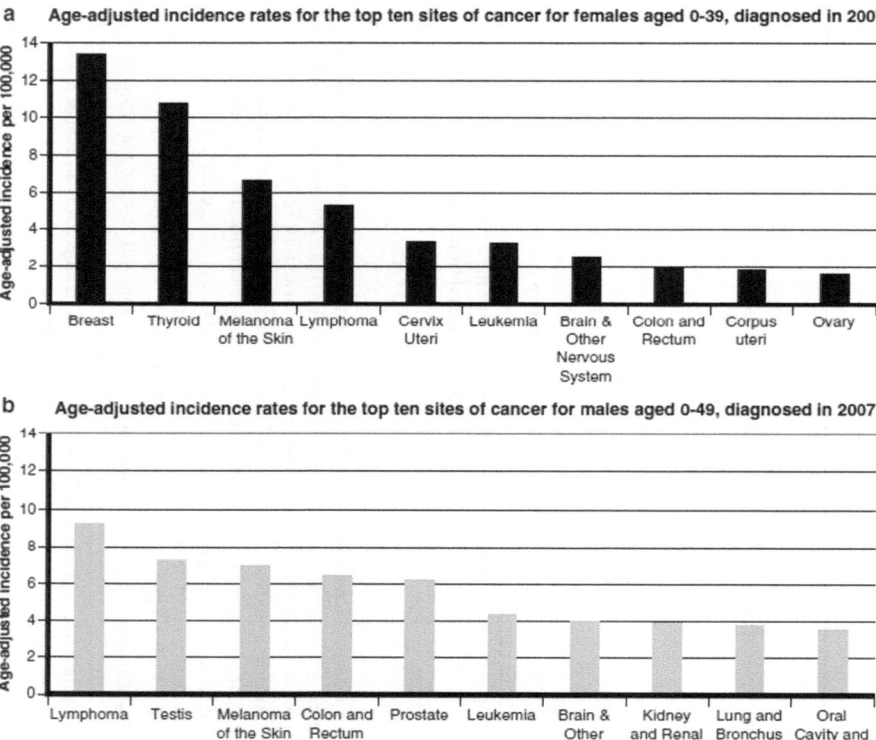

a Age-adjusted incidence rates for the top ten sites of cancer for females aged 0–39, diagnosed in 2007

b Age-adjusted incidence rates for the top ten sites of cancer for males aged 0-49, diagnosed in 2007

Fig. 1.1 (**a**) Age-adjusted incidence rates for the top ten sites for female subjects aged 0–39, diagnosed in 2007. (**b**) Age-adjusted incidence rates for the top ten sites of cancer for male subjects aged 0–49, diagnosed in 2007. Rates are for malignant cases and are adjusted to the 2000 US census population (Data from the SEER 9 registries, including San Francisco-Oakland, Connecticut, Detroit, Hawaii, Iowa, New Mexico, Seattle, Utah, and Atlanta [7])

Cancer Survivorship During Reproductive Ages

In the USA from 1999 to 2006, the 5-year survival of women diagnosed at age 0–44 with any site of cancer was 83.7% [2], For men, this was 75.6%. However, survival is highly dependent on site of cancer: by site at this age group, 5-year survival ranged from 21.9 (esophagus) to 99.4% (thyroid). For major cancers of the female genital system, 5-year survival rates were 89.7 (corpus uteri and uterus not otherwise specified), 83.2 (cervix uteri), and 73.20% (ovary). For cancers of the male genital system, these were 96.7 (testis) and 85.3% (prostate). Overall 5-year survivorship has significantly increased over the past several decades, from 67.7% for men and women aged 0–44 in 1975–1977, to 71.6% in 1993–1995, and to 80.5% in 1999–2006. Increases in 5-year survival for women and men during these periods

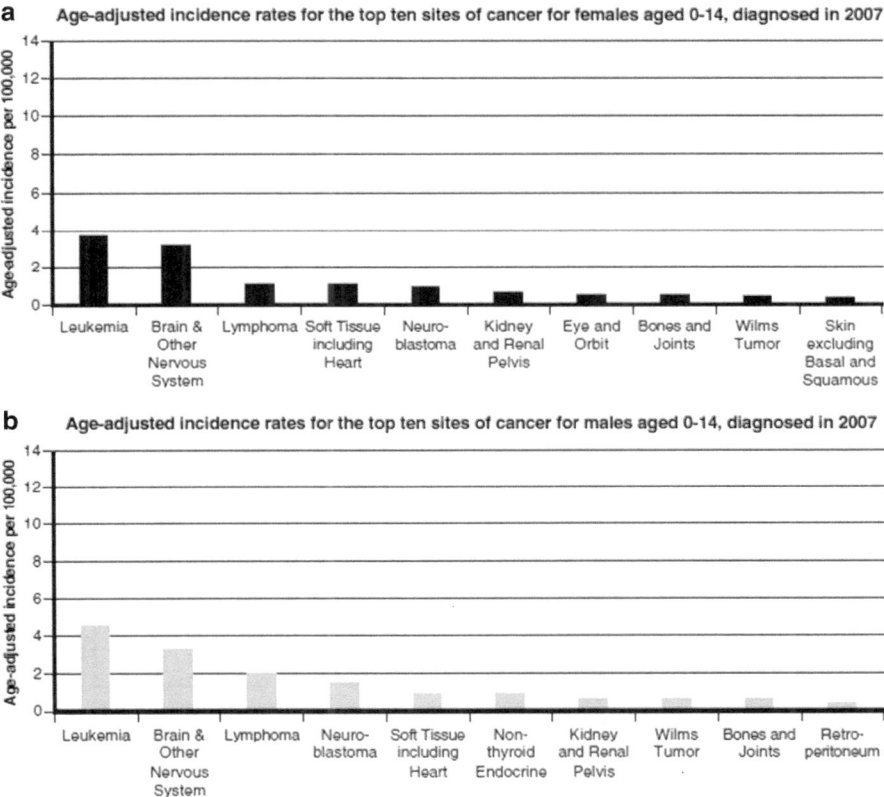

Fig. 1.2 (**a**) Age-adjusted incidence rates for the top ten sites of cancer for female subjects aged 0–14, diagnosed in 2007. (**b**) Age-adjusted incidence rates for the top ten sites of cancer for male subjects aged 0–14, diagnosed in 2007. Rates are for malignant cases and are adjusted to the 2000 US census population (Data from the SEER 9 registries, including San Francisco-Oakland, Connecticut, Detroit, Hawaii, Iowa, New Mexico, Seattle, Utah, and Atlanta [7])

are shown in Fig. 1.3a, b, respectively. Similar increases in cancer survivorship have also been noted in many European countries [8, 9].

For childhood/adolescent cancers in the USA, all-site 5-year survivorship was 81.4%, for cases diagnosed in 1999–2006 [2], compared with 61.7% for cases diagnosed in 1975–1977. Increases in selected sites of cancer for children aged 0–14 are shown in Fig. 1.4. Ten-year survival has also risen, from 58.6% to 75.8% between 1975 and 1997 [2]. A study by Yeh et al. [10] estimated that the overall life expectancy of a 5-year cancer survivor aged 15 was 65.6 years, compared with 76 years for the general US population, representing a loss of 10.4 years. However, this study was based on the data obtained over 20 years ago; life expectancy has likely increased subsequently, due to these improvements in survival rates. It has been estimated that there were 328,652 survivors of childhood cancer in the USA in 2005, of whom 27% were over the age of 39 [11].

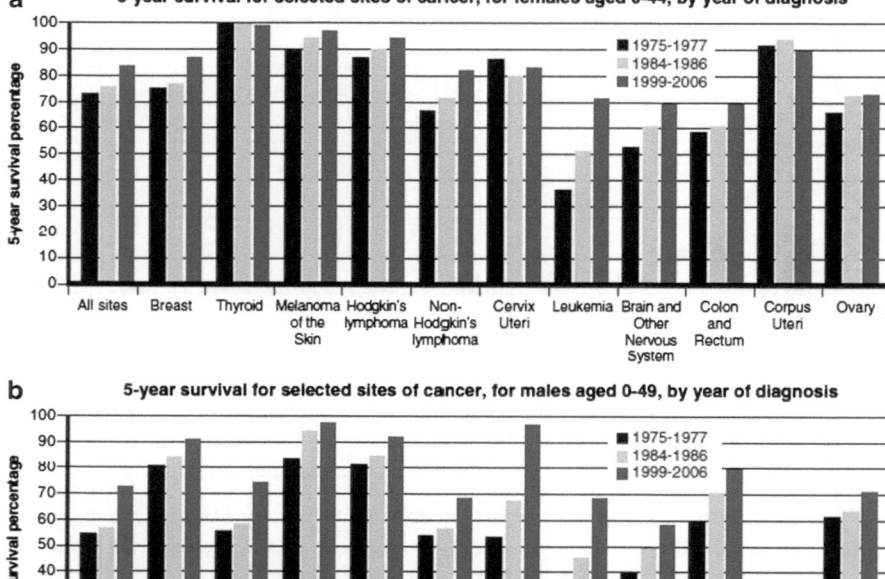

Fig. 1.3 (a) Five-year survival for selected sites of cancer for female subjects aged 0–44 by year of diagnosis. (b) Five-year survival for selected sites of cancer for male subjects aged 0–49 by year of diagnosis (Data from the SEER 9 registries, including San Francisco-Oakland, Connecticut, Detroit, Hawaii, Iowa, New Mexico, Seattle, Utah, and Atlanta [7])

Lupus Survivorship

In addition to cancer, other diseases may also be treated using therapies that compromise fertility. A common treatment for systemic lupus erythematosus (SLE) is the alkylating agent cyclophosphamide, which is known to cause ovarian failure and infertility in patients with this disease [12, 13]. Survival for all forms of lupus has increased substantially, from a 5-year survival of 49% in 1953–1969 to 92% in 1990–1995 [14]. Today, 5-year survival rates are nearly 100%, and 10-year survival rates are almost 90% [15, 16].

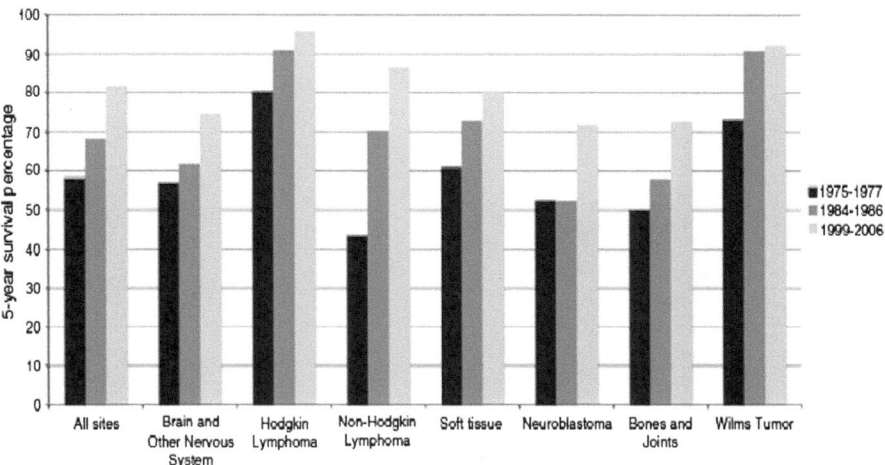

Fig. 1.4 Five-year survival for selected sites of cancer, for boys and girls combined, aged 0–14, by period of diagnosis (Data from the SEER 9 registries, including San Francisco–Oakland, Connecticut, Detroit, Hawaii, Iowa, New Mexico, Seattle, Utah, and Atlanta [2])

Age of First Pregnancy

Among industrialized societies, various factors including prolonged education, career emphasis, longer life span, and availability of effective contraception have contributed toward an increasing age of first pregnancy. From 1970 to 2006, the average age of first pregnancy in the USA increased from 21.4 to 25.0 years [17]. Total birth rates and first birth rates have declined among women aged 25–29 years between 1990 and 2008, but have increased in all subsequent age groups (Fig. 1.5). One percent of first births were from women aged 35 or over in 1970, while this proportion had risen to 8.3 % in 2006 [17, 18]. Although an increase in age of first pregnancy has been observed in all US states, there exists significant interstate variation, with a less than 2.5-year age increase between 1970 and 2006 in New Mexico, Mississippi, and Oklahoma, to a high of 5.0 years or more in Massachusetts, New Hampshire, and the District of Columbia [17]. Similar secular changes in age of first birth have been observed in other developed countries, including most European countries, Canada, and Japan (Fig. 1.6) [17, 97–101]. It has also been suggested that the delay in the age of first birth may be accompanied by a reduction in the inter-pregnancy interval (e.g., time between first and second birth), highlighting the increased urgency of child-bearing toward the end of reproductive age [19].

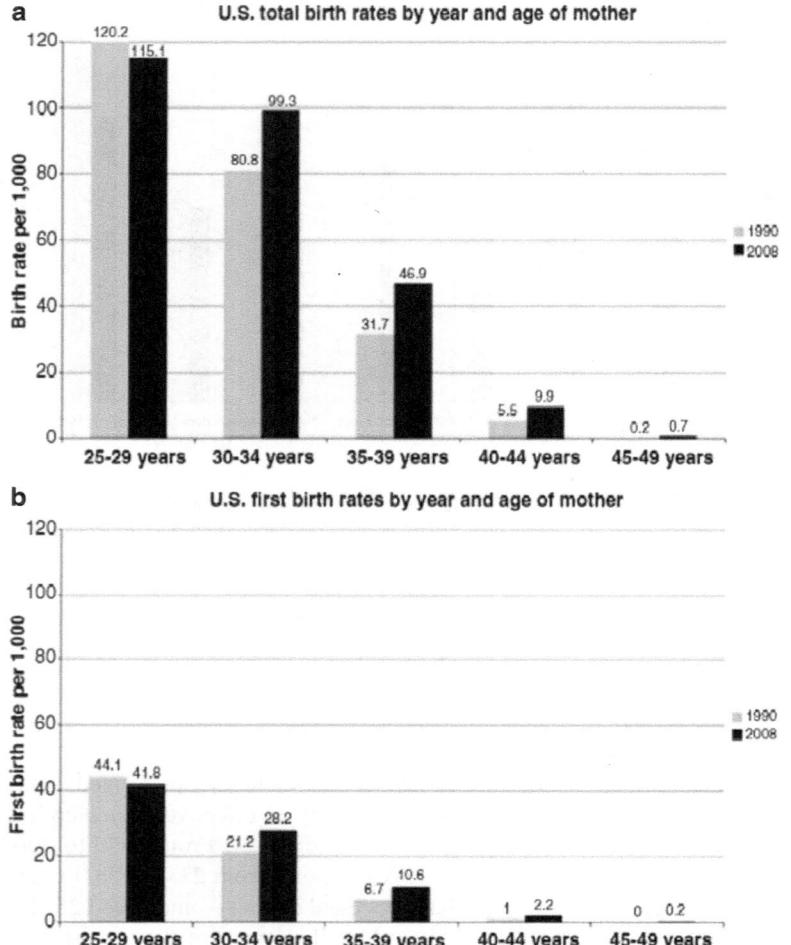

Fig. 1.5 (**a**) US total birth rates by year and age of mother. (**b**) US first birth rates by year and age of mother (Data from US National Vital Statistics Reports [94–96])

Factors that Affect the Fertility of Cancer Survivors: An Overview

The risk and severity of infertility after cancer treatment are highly variable and depend on numerous patient and treatment factors. Patient factors include age of diagnosis, time since treatment, sex, pretreatment fertility, and site and stage of cancer, while treatment factors include type of drug, route of administration, site and size of radiation treatment, dose, and intensity of dose [20]. Treatment of several cancers may require sterilizing surgery. In addition, nonsterilizing surgery

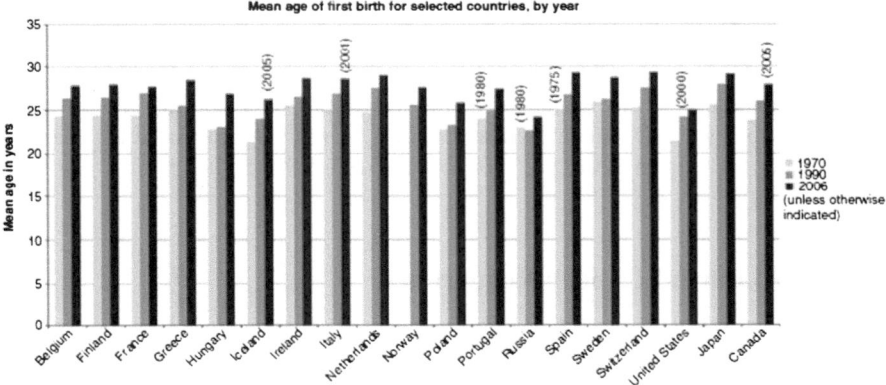

Fig. 1.6 Mean age of first birth for selected countries by year (Data from [17, 97–101])

for cancers of the bladder, prostate, and rectum may also affect fertility by compromising potency and ejaculation [21]. Chemotherapy, and particularly alkylating agents such as cyclophosphamide and procarbazine, can significantly impact fertility, although there are clear dose-, drug-, gender- and age-dependent effects [20, 21]. Radiation at or near reproductive tissues and the pituitary gland [which produces reproductive hormones including follicle-stimulating hormone (FSH) and luteinizing hormone (LH)], or whole body radiation, may also strongly influence fertility [20, 21].

A large cohort study originating from the Childhood Cancer Survivor Study (CCSS) examined predictors of ever siring a pregnancy among 6,224 male subjects without surgical sterility aged 15–44, who were diagnosed before the age of 21 between 1970 and 1986 and survived for at least 5 years after diagnosis [3]. This study found that the likelihood of men ever siring a pregnancy was related to alkylating agent dose, hypothalamic/pituitary radiation dose, testicular radiation dose, type of cancer, and type of chemotherapeutic agent [3]. Subjects who had no exposure to alkylating agents or hypothalamic/pituitary/testes radiation had a likelihood of siring a pregnancy that was not significantly different from their control siblings [hazard ratio (HR): 0.91, 95 % confidence interval (CI): 0.73–1.14]. Increasing doses of each of these exposures were clearly related to a decreasing likelihood of siring a pregnancy, while a significant variation of effect was observed depending on the type of chemotherapeutic agent. Younger age of diagnosis appeared to increase fertility, as those diagnosed between the age of 0 and 4 had an HR of siring of 1.80 (95 % CI: 1.31–2.47), compared with those diagnosed between the age of 15 and 20.

Factors affecting fertility of female cancer survivors were also assessed in the CCSS, using 5,149 subjects without surgical sterility [4]. It was found that the likelihood of ever being pregnant 5 or more years after diagnosis was related to the increasing dose of ovarian/uterine and hypothalamic/pituitary radiation, alkylating agent dose, and type of chemotherapeutic agent. The highest dose of radiation was associated with a relative risk (RR) of infertility of 0.20 (95 % CI: 0.14–0.29, for >10 Gy) for ovarian/uterine exposure, and 0.61 (95 % CI: 0.44–0.85, for >30 Gy) for hypothalamic/pituitary exposure. The highest alkylating agent dose showed a RR of 0.76 (95 % CI: 0.49–0.19). As with men, younger ages had an increased fertility, with those diagnosed between the age of 0 and 4 having an HR of pregnancy of 1.95 (95 % CI: 1.51–2.54).

Risk of Maternal Cancer Recurrence

When considering fertility after cancer, patients and providers must discuss the possibility of recurrence of the cancer. For estrogen-associated cancers in particular, such as those of the breast, it has been hypothesized that pregnancy may promote recurrence via elevated levels of ovarian estrogens that stimulate estrogen-sensitive cell growth and survival. However, studies to date have overwhelmingly found that, for women with breast cancer who wait at least 6 months before attempting to conceive, pregnancy does not affect recurrence or survival compared with women who do not conceive [22–33]. In most cases, pregnancy has been shown to have a protective effect on breast cancer survival. This has been largely attributed to a spurious "healthy mother effect" whereby women who had a better prognosis of cancer were more likely to attempt to conceive, although it may also be due to a true protective effect of hormonal changes during pregnancy [22, 26, 34]. Breast cancer survivors may be advised to wait at least 2 years after treatment before attempting to conceive, to allow adjuvant therapy to be completed and to allow early recurrences to appear [22]. However, conflicting evidence exists regarding the effect of time after diagnosis on survival: while some studies report that time after diagnosis does not modify risk [34–36, 38], others report that the outcome is worse if conception is attempted prior to 6 months after diagnosis of breast cancer [31, 35, 37]. One study has found that, for women with a good breast cancer prognosis, there is a low risk of recurrence for all conception times after diagnosis, while for women with a poor prognosis, there is a high risk of occurrence up to, but not beyond, a wait time of 5 years before conception [41]. Information regarding pregnancy-associated recurrence of other sites of cancer is sparse. Two studies that have examined melanoma have found that pregnancy does not influence survival or recurrence [42, 43].

Health Risks for Offspring of Cancer Survivors

Perinatal Risks

Current evidence suggests that women exposed to abdominal radiation, but not chemotherapy, are at increased risk of pregnancy and neonatal complications. Numerous studies have suggested that abdominal radiation increases the risk of preterm delivery [37–43] and impaired fetal growth [37, 39, 40, 42–48]. Less frequently reported are increased risk for miscarriage [45, 49, 50], postpartum hemorrhage [38], and fetal malposition [44]. Elevated rates of perinatal mortality have also been reported [46, 55], but conflicting evidence has been found [52, 56, 57]. It has been hypothesized that these outcomes are due to radiation-induced damage to uterine tissues, resulting in reduced uterine elasticity and volume [58] and impairment of uterine vasculature remodeling during implantation [53]. Of studies that have examined congenital malformations, most have reported that offspring of cancer survivors have no increased risk [50, 59, 60], although contrary reports exist [46].

The Childhood Cancer Survivor Study found that female survivors of childhood cancer had an overall odds ratio (OR) of preterm delivery of 1.9 (95 % CI: 1.4–2.4), compared with the survivors' siblings [49]. Increasing dose of radiation to the uterus was associated with the increased risk of preterm birth, particularly among subjects who received radiation treatment before menarche; this latter affect was postulated to be due to a higher susceptibility of the immature uterus to radiation damage. Subjects with high doses of radiation to the uterus (250 cGy and above) showed an increased risk for low birth weight, but this was attributed to being born early rather than being an independent outcome. Only uterine radiation doses above 500 cGy showed a significantly increased risk of small-for-gestational-age (SGA). Radiation exposure to the ovaries or to the pituitary showed no significant effect on preterm delivery or low birth weight. Exposure to elevated doses of alkylating agents showed an increased but nonsignificant risk of preterm delivery; no effect was seen for low birth weight or SGA. Similarly, another large study, the British Childhood Cancer Survivor Study found that abdominal radiotherapy increased the risk of preterm delivery (OR 3.2, 95 % CI: 2.1–4.7) and low birth weight (OR, 1.9, 95 % CI: 1.1–3.2), while no significant effects were seen for chemotherapy [47].

Germline Risks

For nonsterile cancer survivors who do not use cryopreserved sperm or embryos, radio- and chemotherapy raise the concern of treatment-related germline mutations affecting the health of offspring. Such germline mutations could theoretically manifest in the form of increased risk of malignancy and genetic anomalies in the child. This can be a common concern for cancer survivors and a potential barrier to seeking parenthood [61, 62], an anxiety that may be emphasized by evidence suggesting that cancer survivors are indeed susceptible to treatment-induced *somatic* mutations that increase the risk of subsequent malignancies [63–65].

However, evidence to date suggests that there is no increased risk of cancer in the children of cancer survivors, when cases with hereditary cancer syndromes are excluded. A population-based study in Finland of cancer patients diagnosed before the age of 35 found that children born more than 9 months after their parent's diagnosis showed no increased risk of cancer, compared with children of the cancer survivors' siblings, after removing individuals with hereditary cancer syndromes [incidence rate ratio (IRR) 1.03, 95 % CI: 0.74–1.40, for all sites of cancer; median follow-up time, 14.9 years since birth] [59]. Moreover, no significant increased risks were observed when stratified by site of the cancer in the child. However, as expected, there was an increased risk of cancer if hereditary cancer cases were not removed (IRR 1.67, 95 % CI: 1.29–2.12). Similarly, a population-based study in Nordic countries from 1943 to 1994 found a nonsignificant risk of all sites of cancer among offspring of cancer survivors (IRR 1.3, 95 % CI: 0.8–2.0; median follow-up time, 14 years since birth), after removing likely cases of hereditary cancers [60]. The same was true when stratified by gender and age at diagnosis. The offspring of almost all subjects in this study were born at least 8 years after cancer was diagnosed in the mothers, and included follow-up to the age of 43 of the offspring. Earlier studies have likewise shown that any increased risk of cancer among offspring is due to familial aggregation rather than mutagenic effects, although these studies were limited by small case numbers and shorter follow-up time [49, 61–63].

Other manifestations of transgenerational genetic effects do not appear to be increased in offspring of cancer survivors, including no differences in cytogenetic syndromes such as Turner's and Down's syndromes, single-gene defects (Mendelian disorders), or genomic instability [64–67], Finally, most studies have reported that there are no significant differences in the gender ratio of either male or female cancer survivors [45, 68–70]. Although one large study did find a significantly altered ratio [71], the authors postulated that this difference was not due to the increased incidence of lethal X chromosome mutations, but rather due to the lowered testosterone levels. Experiments in mice have suggested that chemotherapeutic agents including cyclophosphamide may induce aneuploidy in oocytes, early embryonic mortality, and fetal malformation, but it was postulated that oocytes maturing at the time of exposure are most susceptible to these risks, and that a sufficient interval time between exposure and pregnancy reduces these risks [72]. Chemotherapy has also been associated with aneuploidy in human spermatozoa, but these effects were also found to be transient, lasting for less than 100 days after chemotherapy [73].

The overall health of offspring born to cancer survivors does not appear to be different from expected. A population-based study in Denmark of offspring born between 1977 and 2003 found that children of cancer survivors showed no increased risk of hospitalization compared with the control population (HR 1.05, 95 % CI: 0.98–1.12), after a median follow-up time to age 9.6 [74]. This was true for all discharge diagnoses, including discharges related to infections, all organ and metabolic systems, mental and behavioral problems, and injuries. There was an increased risk of malignant (HR 5.7, 95 % CI: 3.0–10.8) and benign neoplasms (HR 2.0, 95 % CI: 1.2–3.3), but this was attributed to hereditary cancers and increased surveillance, respectively.

Patient Awareness of Fertility Preservation Options

Patients can have significant knowledge deficits regarding options for fertility preservation and the impact of treatment on future fertility, which may be due to emotional stress, urgency of treatment, and a failure by providers to adequately discuss the issue.

A systematic review of the literature using reports from 1999 to 2008 found that, among surveys addressing the issue, between 34 and 72 % of patients are able to recall that they were counseled about the effect of treatment on fertility [5]. Other studies not included in this review have reported recall rates as high as 86 % [75, 76].

Factors related to increased patient knowledge or recall of counseling include having treatment that was more likely to affect fertility, younger age, early stage of disease, participation in decision making, and lower anxiety [75, 77, 78]. Cancers located outside of the reproductive system and treatment without hormone therapy have been associated with the lack of patient knowledge [75]. Women may be less informed than men, possibly due to a belief by medical providers that cryopreservation methods for women are not as easy or effective as those for men [75].

Although the preceding studies are small patient surveys that are highly suscep- tible to selection bias (e.g., patients interested in fertility issues are more likely to respond to surveys), they do emphasize the fact that many patients have incomplete knowledge about fertility at the time of treatment, and that providers selectively provide this knowledge depending on their own beliefs and on characteristics of the patient. For childhood cancer survivors, parents may actively shield their children from learning of the consequences of treatment on future fertility, to avoid increas- ing the emotional burden of their children [55]. It has been noted that the lack of knowledge about fertility status may adversely affect the behavioral decisions of patients, such as whether or not to use contraception [55].

Attitudes and Awareness of Providers

Medical providers play a key role in patient awareness, decision making, and care- seeking behavior regarding fertility preservation. Patients may not raise the issue due to uncertainty or a lack of knowledge, if providers do not take the initiative in starting this conversation [79]. Surveys of physicians and nurses who are involved in the care of cancer patients report that most providers agree that fertility is an important topic worth discussing with their patients, and that general awareness of the topic is high [80–84]. A survey of oncologists, hematologists, and urologists found there was broad agreement that cryopreservation of sperm could help patients psychologically [85]. A majority of physicians report that they routinely discuss the possibility of fertility loss with their patients, particularly among patients deemed to be at high or medium risk of infertility [80–82, 86]. Oncology nurses are less likely to discuss fertility issues with their patients, even though they may believe that it is within their scope of practice [87, 88].

Although fertility may be considered to be an important topic, it is often deemed to be of low priority, both by individual physicians and medical institutions [82, 85], and providers may believe that their patients are more concerned about the topic than they are [96]. One study noted that there exists an important distinction between *discussing* and *mentioning* fertility loss, and that providers may be more likely to do the latter, which may have a significant effect on how the patient con- siders the issue [89].

Even if providers routinely discuss the consequences of fertility with their patients, actual referral rates for fertility preservation may be low. Several surveys have found that a large proportion of physicians do not routinely refer their patients for fertility preservation [79, 82, 84, 85]. A Dutch study conducted in 2002 reported that only 2 % of female patients were referred for fertility preservation [90]. Although this low rate is likely no longer representative given subsequent improvements in cryopreservation options for women, recent studies have likewise suggested that referrals are more common for males than females [80, 81]. Moreover, postpubertal boys are more likely to be referred than early pubertal patients [80]. Some studies have suggested that female physicians are more likely to provide referrals than male physicians [79].

Disease characteristics that are commonly cited by providers to influence decisions over whether to discuss or refer for fertility preservation include urgency of treatment, disease stage, cancer site, age at diagnosis, and treatment type [82, 84, 85, 87, 89]. Additional factors cited by providers that affect likelihood of discussion include level of provider knowledge of fertility preservation options, patient HIV disease status, number of prior children, perception of high cost of fertility services, lack of convenient facilities, marital/relationship status of the patient, patient sexuality, patient interest in fertility preservation, and availability of educational materials [81, 82, 84, 87, 88]. Some providers did not refer because they believed that patients did not assign a high priority to fertility preservation [82, 86], or because they believed that success rates were too low [81]. Difficulty in accommodating parental wishes over whether or not to discuss fertility preservation may also be a factor for pediatric patients [88, 91]. Surveys among nurses have also mentioned that provider's comfort level, including concerns over whether or not adolescent boys should be provided with erotic materials during semen collection, may play a role [87, 88].

Providers may lack the knowledge and training to adequately address the issue of fertility preservation. Numerous surveys have reported that oncology physicians and nurses lack knowledge regarding fertility preservation options, costs, resources, technology, efficacy, and safety, which may lead to underutilization of such services [81, 82, 84, 85, 89]. Seniority may be a factor in awareness of these issues, as more senior physicians may have less training and be less knowledgeable [82]. Physicians may have no formal training in fertility preservation, apart from observing attendings during their fellowship [82]. For training that is provided, it may only involve organ preservation methods rather than cryopreservation [82]. Lack of knowledge may also extend to awareness of the impact of cancer treatment on fertility, as one study found that only 36–62 % of respondents were able to correctly ascribe the risk of gonadotoxicity for several hypothetical cases as based on American Society of Clinical Oncology guidelines [86].

While most studies cited above were surveys conducted among small numbers of physicians and nurses, they point toward a general trend that patient needs for fertility preservation are being incompletely met, and that providers are not adequately prepared or motivated to deal with these issues. With recent advances in fertility preservation methods and increasing focus on long-term quality of life for cancer survivors, this situation is likely to improve in the near future. In 2006, the American Society of Clinical Oncology published guidelines recommending that oncologists should discuss fertility preservation with reproductive age patients, and provide referrals for reproductive services if necessary [20]. Although this will help standardize practice in this area, it has been noted that physician adherence to guidelines may be compromised if physicians do not believe in the efficacy of fertility preservation [82]. Thus, the development of educational materials and training programs for physicians will be important in ensuring that patient needs are met. Social workers and nurses may also be suitable recipients of such training, as it has been suggested that they are in an ideal position to address fertility preservation needs with patients [87, 92, 93].

Finally, providers may need ethical training regarding issues surrounding the recommendation and provisioning of fertility preservation services. Expectations over reduced life span should not be a reason for denying cancer survivors access to fertility services [83]. Similarly, factors such as patient HIV status, number of prior children, marital status, sexuality, and socioeconomic status should not be determinants of whether or not a provider discusses or refers for fertility issues.

Conclusions

Cancer survivors are living longer lives in ever-greater numbers, due to significant improvements in cancer therapy and changes in population age structures. Thus, the focus from merely living to living well has highlighted the need for fertility preserving methods during cancer treatment. Pregnancy for women with a previous diagnosis of cancer is safe, with no evidence of increased risk of recurrence. However, offspring of such women may be susceptible to preterm delivery, low birth weight, and other perinatal morbidities. Both patients and providers alike may have poor knowledge about the fertility preservation during cancer, which highlights the need for increased education in this area.

References

1. Jemal A, Siegel R, Ward E, et al. Cancer statistics, 2007. CA Cancer J Clin. 2007;57:43–66.
2. Seer cancer statistics review 1975–2007.
3. Green DM, Kawashima T, Stovall M, Leisenring W, Sklar CA, Mertens AC, et al. Fertility of male survivors of childhood cancer: a report from the childhood cancer survivor study. J Clin Oncol. 2010;28:332–9.
4. Green DM, Kawashima T, Stovall M, Leisenring W, Sklar CA, Mertens AC, et al. Fertility of female survivors of childhood cancer: a report from the childhood cancer survivor study. J Clin Oncol. 2009;27: 2677–85.
5. Tschudin S, Bitzer J. Psychological aspects of fertility preservation in men and women affected by cancer and other life-threatening diseases. Hum Reprod Update. 2009;15: 587–97.

6. Probability of developing or dying of cancer software, version 6.4.1. Statistical research and applications branch, national cancer institute, 2009. http://srab.Cancer.Gov/devcan

7. Surveillance, epidemiology, and end results (seer) program (www.Seer.Cancer.Gov) seer*stat database: Nov 2009 sub (1973–2007), national cancer institute, dccps, surveillance research program, cancer statistics branch, released April 2010, based on the November 2009 submission.

8. Berrino F, De Angelis R, Sant M, Rosso S, Bielska-Lasota M, Coebergh JW, et al. Survival for eight major cancers and all cancers combined for European adults diagnosed in 1995–99: results of the eurocare-4 study. Lancet Oncol. 2007;8:773–83.

9. Karim-Kos HE, de Vries E, Soerjomataram I, Lemmens V, Siesling S, Coebergh JW. Recent trends of cancer in Europe: a combined approach of incidence, survival and mortality for 17 cancer sites since the 1990s. Eur J Cancer. 2008;44:1345–89.

10. Yeh JM, Nekhlyudov L, Goldie SJ, Mertens AC, Diller L. A model-based estimate of cumulative excess mortality in survivors of childhood cancer. Ann Intern Med. 2010;152:409–17, W131–408.

11. Mariotto AB, Rowland JH, Yabroff KR, Scoppa S, Hachey M, Ries L, et al. Long-term survivors of childhood cancers in the united states. Cancer Epidemiol Biomarkers Prev. 2009; 18:1033–40.

12. Katsifis GE, Tzioufas AG. Ovarian failure in systemic lupus erythematosus patients treated with pulsed intravenous cyclophosphamide. Lupus. 2004; 13:673–8.

13. Langevitz P, Klein L, Pras M, Many A. The effect of cyclophosphamide pulses on fertility in patients with lupus nephritis. Am J Reprod Immunol. 1992;28: 157–8.

14. Cameron JS. Lupus nephritis. J Am Soc Nephrol. 1999;10:413–24.

15. Nossent J, Kiss E, Rozman B, Pokorny G, Vlachoyiannopoulos P, Olesinska M, et al. Disease activity and damage accrual during the early disease course in a multinational inception cohort of patients with systemic lupus erythematosus. Lupus. 2010;19: 949–56.

16. Ravelli A, Ruperto N, Martini A. Outcome in juvenile onset systemic lupus erythematosus. Curr Opin Rheumatol. 2005;17:568–73.

17. Matthews TJ, Hamilton BE. Delayed childbearing: more women are having their first child later in life. NCHS Data Brief 2009;(21):1–8.

18. Mathews TJ, Hamilton BE. Mean age of mother, 1970–2000. Natl Vital Stat Rep. 2002; 51(1):1–13.

19. Kalberer U, Baud D, Fontanet A, Hohlfeld P, de Ziegler D. Birth records from Swiss married couples analyzed over the past 35 years reveal an aging of first-time mothers by 5.1 years while the interpregnancy interval has shortened. Fertil Steril. 2009;92:2072–3.

20. Lee SJ, Schover LR, Partridge AH, Patrizio P, Wallace WH, Hagerty K, et al. American society of clinical oncology recommendations on fertility preservation in cancer patients. J Clin Oncol. 2006;24: 2917–31.

21. Dohle GR. Male infertility in cancer patients: review of the literature. Int J Urol. 2010; 17:327–31.

22. de Bree E, Makrigiannakis A, Askoxylakis J, Melissas J, Tsiftsis DD. Pregnancy after breast cancer: a comprehensive review. J Surg Oncol. 2010;101: 534–42.

23. Blakely LJ, Buzdar AU, Lozada JA, Shullaih SA, Hoy E, Smith TL, et al. Effects of pregnancy after treatment for breast carcinoma on survival and risk of recurrence. Cancer. 2004;100:465–9.

24. Clark RM, Chua T. Breast cancer and pregnancy: the ultimate challenge. Clin Oncol (R Coll Radiol). 1989;1:11–8.

25. Danforth Jr DN. How subsequent pregnancy affects outcome in women with a prior breast cancer. Oncology (Williston Park). 1991;5:23–30. discussion 30–21, 35.

26. Gelber S, Coates AS, Goldhirsch A, Castiglione-Gertsch M, Marini G, Lindtner J, et al. Effect of pregnancy on overall survival after the diagnosis of early-stage breast cancer. J Clin Oncol. 2001; 19: 1671–5.

27. Harvey JC, Rosen PP, Ashikari R, Robbins GF, Kinne DW. The effect of pregnancy on the prognosis of carcinoma of the breast following radical mastectomy. Surg Gynecol Obstet. 1981;153:723–5.

28. Ives A, Saunders C, Bulsara M, Semmens J. Pregnancy after breast cancer: Population based study. BMJ. 2007;334:194.
29. Mignot L, Morvan F, Berdah J, Querleu D, Laurent JC, Verhaeghe M, et al. [Pregnancy after treated breast cancer. Results of a case–control study]. Presse Med. 1986;15:1961–4.
30. Mueller BA, Simon MS, Deapen D, Kamineni A, Malone KE, Daling JR. Childbearing and survival after breast carcinoma in young women. Cancer. 2003;98:1131–40.
31. Sankila R, Heinavaara S, Hakulinen T. Survival of breast cancer patients after subsequent term pregnancy: "Healthy mother effect". Am J Obstet Gynecol. 1994;170:818–23.
32. Velentgas P, Daling JR, Malone KE, Weiss NS, Williams MA, Self SG, et al. Pregnancy after breast carcinoma: outcomes and influence on mortality. Cancer. 1999;85:2424–32.
33. Weisz B, Schiff E, Lishner M. Cancer in pregnancy: maternal and fetal implications. Hum Reprod Update. 2001;7:384–93.
34. Largillier R, Savignoni A, Gligorov J, Chollet P, Guilhaume MN, Spielmann M, et al. Prognostic role of pregnancy occurring before or after treatment of early breast cancer patients aged <35 years: a get(n)a working group analysis. Cancer. 2009;115:5155–65.
35. Grin CM, Driscoll MS, Grant-Kels JM. The relationship of pregnancy, hormones, and melanoma. Semin Cutan Med Surg. 1998;17:167–71.
36. Reintgen DS, McCarty Jr KS, Vollmer R, Cox E, Seigler HF. Malignant melanoma and pregnancy. Cancer. 1985;55:1340–4.
37. Green DM, Peabody EM, Nan B, Peterson S, Kalapurakal JA, Breslow NE. Pregnancy outcome after treatment for wilms tumor: a report from the national wilms tumor study group. J Clin Oncol. 2002;20:2506–13.
38. Lie Fong S, van den Heuvel-Eibrink MM, Eijkemans MJ, Schipper I, Hukkelhoven CW, Laven JS. Pregnancy outcome in female childhood cancer survivors. Hum Reprod. 2010;25: 1206–12.
39. Magelssen H, Melve KK, Skjaerven R, Fossa SD. Parenthood probability and pregnancy outcome in patients with a cancer diagnosis during adolescence and young adulthood. Hum Reprod. 2008;23: 178–86.
40. Reulen RC, Zeegers MP, Wallace WH, Frobisher C, Taylor AJ, Lancashire ER, et al. Pregnancy outcomes among adult survivors of childhood cancer in the British childhood cancer survivor study. Cancer Epidemiol Biomarkers Prev. 2009;18:2239–47.
41. Sanders JE, Hawley J, Levy W, Gooley T, Buckner CD, Deeg HJ, et al. Pregnancies following high-dose cyclophosphamide with or without high-dose busulfan or total-body irradiation and bone marrow transplantation. Blood. 1996;87:3045–52.
42. Signorello LB, Cohen SS, Bosetti C, Stovall M, Kasper CE, Weathers RE, et al. Female survivors of childhood cancer: preterm birth and low birth weight among their children. J Natl Cancer Inst. 2006;98: 1453–61.
43. Sudour H, Chastagner P, Claude L, Desandes E, Klein M, Carrie C, et al. Fertility and pregnancy outcome after abdominal irradiation that included or excluded the pelvis in childhood tumor survivors. Int J Radiat Oncol Biol Phys. 2010;76:867–73.
44. Chiarelli AM, Marrett LD, Darlington GA. Pregnancy outcomes in females after treatment for childhood cancer. Epidemiology. 2000;11:161–6.
45. Green DM, Whitton JA, Stovall M, Mertens AC, Donaldson SS, Ruymann FB, et al. Pregnancy outcome of female survivors of childhood cancer: a report from the childhood cancer survivor study. Am J Obstet Gynecol. 2002;187:1070–80.
46. Hawkins MM, Smith RA. Pregnancy outcomes in childhood cancer survivors: probable effects of abdominal irradiation. Int J Cancer. 1989;43: 399–402.
47. Salooja N, Szydlo RM, Socie G, Rio B, Chatterjee R, Ljungman P, et al. Pregnancy outcomes after peripheral blood or bone marrow transplantation: a retrospective survey. Lancet. 2001;358:271–6.
48. Li FP, Gimbrere K, Gelber RD, Sallan SE, Flamant F, Green DM, et al. Outcome of pregnancy in survivors of wilms' tumor. JAMA. 1987;257:216–9.
49. Hawkins MM, Draper GJ, Smith RA. Cancer among 1,348 offspring of survivors of childhood cancer. Int J Cancer. 1989;43:975–8.

50. Winther JF, Boice Jr JD, Svendsen AL, Frederiksen K, Stovall M, Olsen JH. Spontaneous abortion in a Danish population-based cohort of childhood cancer survivors. J Clin Oncol. 2008;26:4340–6.
51. Critchley HO, Bath LE, Wallace WH. Radiation damage to the uterus –review of the effects of treatment of childhood cancer. Hum Fertil (Camb). 2002;5:61–6.
52. Sawka AM, Lakra DC, Lea J, Alshehri B, Tsang RW, Brierley JD, et al. A systematic review examining the effects of therapeutic radioactive iodine on ovarian function and future pregnancy in female thyroid cancer survivors. Clin Endocrinol (Oxf). 2008;69:479–90.
53. Winther JF, Boice Jr JD, Frederiksen K, Bautz A, Mulvihill JJ, Stovall M, et al. Radiotherapy for childhood cancer and risk for congenital malformations in offspring: a population-based cohort study. Clin Genet. 2009;75:50–6.
54. Langeveld NE, Ubbink MC, Last BF, Grootenhuis MA, Voute PA, De Haan RJ. Educational achievement, employment and living situation in long-term young adult survivors of childhood cancer in the Netherlands. Psychooncology. 2003;12:213–25.
55. Zebrack BJ, Casillas J, Nohr L, Adams H, Zeltzer LK. Fertility issues for young adult survivors of childhood cancer. Psychooncology. 2004;13:689–99.
56. Goldsby R, Burke C, Nagarajan R, Zhou T, Chen Z, Marina N, et al. Second solid malignancies among children, adolescents, and young adults diagnosed with malignant bone tumors after 1976: follow-up of a children's oncology group cohort. Cancer. 2008; 113:2597–604.
57. Meadows AT, Friedman DL, Neglia JP, Mertens AC, Donaldson SS, Stovall M, et al. Second neoplasms in survivors of childhood cancer: findings from the childhood cancer survivor study cohort. J Clin Oncol. 2009;27:2356–62.
58. Sankila R, Pukkala E, Teppo L. Risk of subsequent malignant neoplasms among 470,000 cancer patients in Finland, 1953–1991. Int J Cancer. 1995;60: 464–70.
59. Madanat-Harjuoja LM, Malila N, Lahteenmaki P, Pukkala E, Mulvihill JJ, Boice Jr JD, et al. Risk of cancer among children of cancer patients - a nationwide study in Finland. Int J Cancer. 2010;126: 1196–205.
60. Sankila R, Olsen JH, Anderson H, Garwicz S, Glattre E, Hertz H, et al. Risk of cancer among offspring of childhood-cancer survivors. Association of the nordic cancer registries and the nordic society of paediatric haematology and oncology. N Engl J Med. 1998;338: 1339–44.
61. Bessho F, Kobayashi M. Adult survivors of children's cancer and their offspring. Pediatr Int. 2000;42: 121–5.
62. Hawkins MM, Draper GJ, Winter DL. Cancer in the offspring of survivors of childhood leukaemia and non-hodgkin lymphomas. Br J Cancer. 1995;71: 1335–9.
63. Mulvihill JJ, Myers MH, Connelly RR, Byrne J, Austin DF, Bragg K, et al. Cancer in offspring of long-term survivors of childhood and adolescent cancer. Lancet. 1987;2:813–7.
64. Bajnoczky K, Khezri S, Kajtar P, Szucs R, Kosztolanyi G, Mehes K. No chromosomal instability in offspring of survivors of childhood malignancy. Cancer Genet Cytogenet. 1999;109:79–80.
65. Byrne J, Rasmussen SA, Steinhorn SC, Connelly RR, Myers MH, Lynch CF, et al. Genetic disease in offspring of long-term survivors of childhood and adolescent cancer. Am J Hum Genet. 1998;62:45–52.
66. Tawn EJ, Whitehouse CA, Winther JF, Curwen GB, Rees GS, Stovall M, et al. Chromosome analysis in childhood cancer survivors and their offspring-no evidence for radiotherapy-induced persistent genomic instability. Mutat Res. 2005;583:198–206.
67. Winther JF, Boice Jr JD, Mulvihill JJ, Stovall M, Frederiksen K, Tawn EJ, et al. Chromosomal abnormalities among offspring of childhood-cancer survivors in Denmark: a population-based study. Am J Hum Genet. 2004;74:1282–5.
68. Chow EJ, Kamineni A, Daling JR, Fraser A, Wiggins CL, Mineau GP, et al. Reproductive outcomes in male childhood cancer survivors: a linked cancer-birth registry analysis. Arch Pediatr Adolesc Med. 2009;163: 887–94.
69. Reulen RC, Zeegers MP, Lancashire ER, Winter DL, Hawkins MM. Offspring sex ratio and gonadal irradiation in the British childhood cancer survivor study. Br J Cancer. 2007;96:1439–41.

70. Winther JF, Boice Jr JD, Thomsen BL, Schull WJ, Stovall M, Olsen JH. Sex ratio among off-spring of childhood cancer survivors treated with radiotherapy. Br J Cancer. 2003;88:382–7.
71. Green DM, Whitton JA, Stovall M, Mertens AC, Donaldson SS, Ruymann FB, et al. Pregnancy outcome of partners of male survivors of childhood cancer: a report from the childhood cancer survivor study. J Clin Oncol. 2003;21:716–21.
72. Meirow D, Epstein M, Lewis H, Nugent D, Gosden RG. Administration of cyclophosph-amide at different stages of follicular maturation in mice: effects on reproductive perfor-mance and fetal malformations. Hum Reprod. 2001;16:632–7.
73. Robbins WA, Meistrich ML, Moore D, Hagemeister FB, Weier HU, Cassel MJ, et al. Chemotherapy induces transient sex chromosomal and autosomal aneuploidy in human sperm. Nat Genet. 1997;16:74–8.
74. Winther JF, Boice JD Jr, Christensen J, Frederiksen K, Mulvihill JJ, Stovall M, Olsen JH. Hospitalizations among children of survivors of childhood and adolescent cancer: a popula-tion-based cohort study. Int J Cancer. 2010;127:2879–87.
75. Mancini J, Rey D, Preau M, Malavolti L, Moatti JP. Infertility induced by cancer treatment: inappropriate or no information provided to majority of French survivors of cancer. Fertil Steril. 2008;90:1616–25.
76. Partridge AH, Gelber S, Peppercorn J, Sampson E, Knudsen K, Laufer M, et al. Web-based survey of fertility issues in young women with breast cancer. J Clin Oncol. 2004;22:4174–83.
77. Duffy CM, Allen SM, Clark MA. Discussions regarding reproductive health for young women with breast cancer undergoing chemotherapy. J Clin Oncol. 2005;23:766–73.
78. Schover LR. Psychosocial aspects of infertility and decisions about reproduction in young cancer survivors: a review. Med Pediatr Oncol. 1999;33:53–9.
79. Quinn GP, Vadaparampil ST, Lee JH, Jacobsen PB, Bepler G, Lancaster J, et al. Physician referral for fertility preservation in oncology patients: a national study of practice behaviors. J Clin Oncol. 2009;27: 5952–7.
80. Anderson RA, Weddell A, Spoudeas HA, Douglas C, Shalet SM, Levitt G, et al. Do doctors discuss fertility issues before they treat young patients with cancer? Hum Reprod. 2008;23:2246–51.
81. Goodwin T, Elizabeth Oosterhuis B, Kiernan M, Hudson MM, Dahl GV. Attitudes and prac-tices of pediatric oncology providers regarding fertility issues. Pediatr Blood Cancer. 2007;48:80–5.
82. Quinn GP, Vadaparampil ST, Gwede CK, Miree C, King LM, Clayton HB, et al. Discussion of fertility preservation with newly diagnosed patients: oncologists' views. J Cancer Surviv. 2007;1:146–55.
83. Robertson JA. Cancer and fertility: ethical and legal challenges. J Natl Cancer Inst Monogr 2005;(34): 104–6.
84. Schover LR, Brey K, Lichtin A, Lipshultz LI, Jeha S. Oncologists' attitudes and practices regarding banking sperm before cancer treatment. J Clin Oncol. 2002;20:1890–7.
85. Allen C, Keane D, Harrison RF. A survey of Irish consultants regarding awareness of sperm freezing and assisted reproduction. Ir Med J. 2003;96:23–5.
86. Forman EJ, Anders CK, Behera MA. A nationwide survey of oncologists regarding treat-ment-related infertility and fertility preservation in female cancer patients. Fertil Steril. 2010;94:1652–6.
87. King L, Quinn GP, Vadaparampil ST, Gwede CK, Miree CA, Wilson C, et al. Oncology nurses' perceptions of barriers to discussion of fertility preservation with patients with can-cer. Clin J Oncol Nurs. 2008;12:467–76.
88. Vadaparampil ST, Clayton H, Quinn GP, King LM, Nieder M, Wilson C. Pediatric oncology nurses' attitudes related to discussing fertility preservation with pediatric cancer patients and their families. J Pediatr Oncol Nurs. 2007;24:255–63.
89. Zapzalka DM, Redmon JB, Pryor JL. A survey of oncologists regarding sperm cryopreservation and assisted reproductive techniques for male cancer patients. Cancer. 1999;86:1812–7.
90. Jenninga E, Hilders CG, Louwe LA, Peters AA. Female fertility preservation: practical and ethical considerations of an underused procedure. Cancer J. 2008;14:333–9.

91. de Vries MC, Bresters D, Engberts DP, Wit JM, van Leeuwen E. Attitudes of physicians and parents towards discussing infertility risks and semen cryopreservation with male adolescents diagnosed with cancer. Pediatr Blood Cancer. 2009;53:386–91.
92. Clayton H, Quinn GP, Lee JH, King LM, Miree CA, Nieder M, et al. Trends in clinical practice and nurses' attitudes about fertility preservation for pediatric patients with cancer. Oncol Nurs Forum. 2008;35:249–55.
93. King L, Quinn GP, Vadaparampil ST, Miree CA, Wilson C, Clayton H, et al. Oncology social workers' perceptions of barriers to discussing fertility preservation with cancer patients. Soc Work Health Care. 2008;47:479–501.
94. Hamilton BE, Sutton PD, Ventura SJ. Revised birth and fertility rates for the 1990s and new rates for Hispanic populations, 2000 and 2001: United States. Natl Vital Stat Rep. 2003;51:1–94.
95. Martin JA, Hamilton BE, Sutton PD, Ventura SJ, Menacker F, Kimeyer S, Matthews TJ. Births: final data for 2006. Natl Vital Stat Rep. 2006;57: 1–104.
96. Hamilton BE, Martin JA, Ventura SJ. Births: preliminary data for 2008. Natl Vital Stat Rep. 2010;58:1–17.
97. Vienna Institute of Demography. European demographic data sheet 2008. Retrieved from: http://www.oeaw.ac.at/vid/datasheet/download/sources_notes_datasheet2008.pdf. [Retrieved 7/20/11].
98. Statistics Canada. Report on the demographic situation in Canada, 2005 and 2006. Retrieved from: [Retrieved 7/20/11].
99. Council of Europe Publishing. Recent demographic developments in Europe, 2002. Retrieved from: http://www.coe.int/t/e/social_cohesion/population/d%C3%A9mo211960EN.PDF. [Retrieved 7/20/11].
100. United Nations economic commission for Europe. Trends in Europe and North America. The statistical yearbook of the economic commission for Europe 2003.
101. Mathews TJ, Hamilton BE. Mean age of mother, 1970–2000. Natl Vital Stat Rep. 2002;51:1–13

Chapter 2
Ethical Discussions in Approaching Fertility Preservation

Pasquale Patrizio

Ethical Principles: General Considerations

Before discussing the ethical dilemmas associated with fertility preservation, it is necessary to describe the ethical constructs most commonly used to formulate guidelines. It is also important to realize that any policy or guideline needs to have a certain degree of built-in flexibility, because natural rights and human dignity, in the context of creating families, cannot always be clearly defined. It is understandably difficult to balance an individual right to reproductive autonomy and privacy with the societal obligations to protect the potentiality of life.

The most common ethical norms used to conduct ethical analyses are based on the five autonomy, beneficence, nonmaleficence, justice, and veracidity.

Autonomy

Autonomy or *respect* for persons, acknowledges an individual's right to hold views, makes choices and takes actions on the basis of personal values and beliefs. This principle is at the base of informed consent and respect for privacy. In the reproductive field, this principle is the basis for reproductive rights and reproductive choices. Some examples: Should fertility preservation be offered to a woman who knows that her prognosis for long-term survival is uncertain? Should a widower be permitted to use embryos frozen while his wife was alive, and to use a member of his wife's family as a gestational carrier?

P. Patrizio, M.D., MBE, HCLD(✉)
Yale Fertility Center and REI Medical Practice,
150 Sargent Drive (2nd Floor), New Haven, CT 06511, USA
e-mail: Pasquale.Patrizio@yale.edu

E. Seli and A. Agarwal (eds.), *Fertility Preservation in Males: Emerging Technologies
and Clinical Applications*, DOI 10.1007/978-1-4614-5620-9_2,
© Springer Science+Business Media New York 2012

As to each question posed, the formulation of guidelines must take into account the right of an individual to decide, with the understanding that procreative liberty has some limits when these rights conflict with the child's best interest. In formulating an informed consent, it is therefore important to anticipate these scenarios and request disposal directives for cryopreserved reproductive tissue or gametes or embryos.

Beneficence

Beneficence represents the obligation to promote the patient's well-being. Beneficence represents the balancing of risks and benefits of an intervention. It is the principle that dictates that subjects be protected from harm and that efforts be taken to secure their well-being. It may conflict with the principle of autonomy. For example, the woman with breast cancer that requests controlled ovarian hyperstimulation despite being positive for estrogen and progesterone receptors and carrying breast cancer gene mutations.

Nonmaleficence

Nonmaleficence is the obligation to "do no harm" also known by its original Latin expression of "primum non nocere," which needs to be balanced with the principle of autonomy. Both beneficence and non-maleficence may overlap, for example, a patient who wants to postpone the beginning of chemotherapy treatments and, contrary to the oncologist opinion, insists in undergoing fertility preservation.

Justice

Justice concerns fairness and equity, i.e., the need to be fair in sharing the burden (costs) and the distribution (benefits) of resources to all members of the community. In particular, the concept of *distributive justice* is often applied to situations requiring a decision about the equitable allocation of resources. The current practice of IVF in general and fertility preservation in particular however, is not fair. Since many options to preserve fertility are experimental, insurance companies are not covering these services. Particularly for low income people, these services are only offered on institutional grants or on charitable basis; consequently, most patients are unable to have access to these services.

Veracidity

The principle of *veracidity* stipulates that a provider always tells the truth to his/her patient and avoids the exploitation of vulnerable populations.

Fertility Preservation for Cancer: Ethical Considerations

Ethical Consideration for Fertility Preservation in Adults

Fertility preservation developed first with the intent of preserving the potential for genetic parenthood in adults or children at risk of sterility due to chemo- or radio-therapy. Today, many young patients with cancer are surviving (the 5-year survival rates for Caucasian and Hispanic American women have increased for Hodgkin's lymphoma from 86% to 98% in the quarter century to the year 2000, and for breast cancer from 78% to 91%) [1]. At the same time, diagnoses of some malignant diseases have become more prevalent (e.g., breast and testicular cancer) [2]. The net effect has been an increase in numbers of patients in their reproductive years at risk of sterilization or early menopause [1]. As a result of this progress, quality of life issues after cancer is becoming increasingly more important and protection of fertility is a preeminent quality of life paradigm.

Although there are many strategies to preserve fertility, embryo freezing and sperm freezing are the only established options while all the others are still considered experimental; experimental procedures include oocyte freezing, ovarian tissue or whole ovary freezing, and in vitro maturation of oocytes or in vitro folliculogenesis.

Likewise, for men, when the option of semen cryopreservation is not available as for prepubertal boys, the harvesting and isolation of spermatogonial cells from testicular biopsies or the freezing of testicular tissue for later transplantation or even xenografting, are being tested but remain highly experimental.

When using experimental techniques, the informed consent is essential and both women and men have the right to know all options concerning fertility preservation and their implications including the risks and costs involved. In addition, experimental procedures are considered under the umbrella of research protocols and thus should also be reviewed and approved by institutional review boards.

Providing thorough informed consent in recruiting persons to participate in research is the foundation for the ethical conduct of research. It is based on three components: adequate, comprehensible information; a competent decision-maker; and a voluntary decision process.

Patients have the right to know what will happen if they or any children that are created are injured or disabled, including issues related to health insurance and compensation. Providers obtaining consents for experimental techniques must focus not simply on disclosure but also on effective communication and comprehension. The use of quizzes and documenting responses to questions after information is presented are effective tools to assist in demonstrating that patients understand the experimental or innovative nature of some modes of fertility preservation. Consent to the use of one of the many therapeutic strategies may require involving a surrogate decision maker in the case of young children or mentally impaired persons.

From an ethical standpoint, the key reason for pursuing fertility preservation is to restore personal autonomy to those who might, in the future, become unable to conceive [3, 4]. The presentation of risk information is complicated by the fact that both the adult and their offspring may be involved. A core principle of medical ethics is to do no harm. Ideally, the decision about who is candidate for fertility preservation should be rendered by a team including a medical oncologist, a reproductive endocrinologist, a pathologist and a psychologist, all guided by written protocols that can be shared with patients [4]. Patients should not be provided with false hopes. Alternative plans including no intervention with the prospect of adoption or childlessness should also be a part of the discussion. Equity or ownership interests of caregivers in novel technologies utilized in research must be disclosed to potential subjects. It is reasonable in the absence of grant funds to seek reimbursement from patients to cover the expenses of the research, but there should be no charge for clinical fees until the experimental options have been proven safe and effective.

Concerns about the children of mothers affected by cancer fall into three categories: first, the possible shortened life expectancy of mothers should cancer recur meaning that children could be orphaned at an early age. Another is the health of the mothers' posttreatment and the fear that they will not have the energy and stamina to care for their children. The third is the health of the children born after the thawing and retransplantation of ovarian cortical tissue or from the in vitro maturation of follicles or oocytes. Therefore, in the absence of any long-term follow-up studies or registries, it is imperative that those involved with these techniques continue monitoring each of these issues. It also means that novel forms of fertility preservation involving, for example, ovarian and testicular harvesting for freezing should be performed only in a few specialized centers working with proper IRB permission, the capacity for follow-up with subjects, adequate social service support, and subjects signing ethics committee approved consents.

Ethical Consideration for Fertility Preservation in Children

Impaired future fertility is a possible consequence of exposure to cancer therapies even for children. This risk may be difficult for children to conceptualize, but potentially traumatic to them when they become adults.

Since the modalities that are available to children for preserving their fertility are limited by their sexual immaturity, they are all considered experimental. For prepubertal boys who cannot produce mature sperm, harvesting and cryopreservation of testicular stem cells with the hope of future autologous transplantation, or in vitro maturation, represent potential methods of fertility preservation. For girls, isolation and cryopreservation of ovarian cortical strips/primordial follicles followed by in vitro maturation of gametes, when fertility is desired, is a possible option. Extensive research is still required to refine these modalities to safely offer them to patients as therapies [4].

Again, assisted reproductive technologies must be scrutinized on the basis of efficacy and safety and they must be subjected to rigorous ethical deliberation by independent review board committees before they can be offered. The modalities involved in fertility preservation of young children are no exception to these rules.

In addition to ensuring that the basis for offering the intervention is scientifically sound, the execution of the intervention must be deemed ethically sound. This determination requires that the intervention in question be evaluated within an ethical framework that considers it in terms of beneficence, respect for persons (autonomy), and justice [5].

It can be argued that fertility preservation aimed at children is ethical because it prevents morbidity (reproductive and psychosocial) and it safeguards their reproductive autonomy [6]. Therefore, the main ethical question concerns the process and the techniques necessary to protect fertility. The special situation of children as research subjects and at the same time patients makes the provider open to potential abuse of the technologies in an impetus to have a breakthrough [6, 7]. To avoid this risk, it is prudent to have multiple caregivers involved in the consent process.

Programs must make every effort to minimize financial barriers to access for children and to work with patient advocacy groups to seek coverage for children and families who cannot afford to participate in fertility treatment or research. Research involved in childhood fertility preservation should be conducted on patients who could experience personal benefit from the research, eliminating the prospect of exploitation for the gain of others.

Children represent a unique and vulnerable population with respect to medical research. They have diminished autonomy, diminished capacity to understand the risks and benefits of research objectives and lack the ability to provide consent for research studies. As a result, they require special protection against potential violation of their rights that may occur during research investigations [4, 5]. Until very recently, institutional attitudes impeded significant participation by children in medical research for the fear of exploitation.

Children should not be exploited to participate in pediatric research, nor should they be deprived of the benefits research has to offer because of their vulnerable status. With respect to childhood fertility preservation, proper attainment of informed consent from a legally authorized representative (i.e., parent or guardian) and of childhood assent must be ensured [4, 5]. Assent-the active affirmation by the research subject-can be obtained from incompetent minors and it should be obtained from children whenever possible. While the benefits of gamete cryopreservation are promising, they are largely unquantifiable because human data on the survival of gametes after the freeze-thaw-transplant process are limited. Until more data become available providers cannot tell patients what percentage of gametes will survive and what the probability of conception is, and must not provide patients with false hope. Alternatives to gamete cryopreservation should be discussed and patients should be given the option of no intervention [6]. Barriers to the consent process for fertility preservation interventions may develop. While parents may be competent to consent for their children, the scenario is very complex clinically and emotionally [7].

It has been suggested that to overcome some of the practical obstacles involved in the consent process, it should be performed in stages [8]. If a two-stage process is adopted, the issues of gonadal harvesting/storage and gamete manipulation can be handled as two separate topics at distinct time points. The decision to harvest gametes would be made at the time of cancer diagnosis and consent for the procedure would be left to parents/guardians. The decision of whether to use the gametes after they have been isolated can then be made at a future point by the child, when adulthood is reached. At such a point in time, the young patient would be better able to express personal preferences about the handling of the tissue based on an enhanced capacity to understand the ramifications of the possible medical interventions available at that time.

Summary of Clinical Outcomes Within the Context of Informed Consent

Health of Children Born from Oocyte Cryopreservation

Many patients with cancer are choosing the option of oocyte cryopreservation as a fertility preservation strategy. However, despite the birth of hundreds of children, the ASRM still considers this technology as experimental [9, 10]. Recent summary reports have documented that babies born from the use of cryopreserved oocytes by vitrification methods (outcome of 200 infants born) are not at higher risk for congenital malformations and not at an increased risk of adverse perinatal outcome [9].

Another study summarized 58 reports (1986–2008) that included 609 live born babies (308 from slow freezing, 289 from vitrification, and 12 from both methods) [10]. In addition, 327 other live births were verified. Of the total 936 infants, 1.3% [12] had a birth anomaly: three ventricular septal defects, one choanal and one biliary atresia, one Rubinstein-Taybi syndrome, one Arnold-Chiari syndrome, one cleft palate, three clubfoot, and one skin hemangioma. On the whole, these observations demonstrate that, so far, children conceived from oocyte freezing are healthy and not at an increased risk of adverse outcome.

Health of Children Born from Cryopreserved Embryos

The first prospective study aimed at assessing the postnatal growth and development of children born from cryopreserved embryos was carried out in 1995 [11]. The findings of that study were that children conceived from cryopreservation had a lower mean birth weight and mean gestational age, but the incidence of minor and major congenital malformations was similar to that of a control group and, furthermore, these children performed on a similar functional level as the control group.

A recent systematic review evaluated the medical outcome of children born after cryopreservation, slow freezing and vitrification of early cleavage stage embryos, blastocysts and oocytes during the years 1984–2008 [12]. Most studies found comparable malformation rates between frozen and fresh ART cycles and overall data concerning infant outcome and psychological well-being after cryopreservation of embryos were reassuring. As for oocyte cryopreservation data, the number of properly controlled follow-up studies of neonatal outcome after embryo cryopreservation is still somewhat limited. Long-term follow-up studies for all cryopreservation techniques are also essential.

Health of Children Born from Frozen/Retransplanted Human Ovarian Tissue

Ovarian cryopreservation and transplantation techniques, either as heterotopic or orthotopic allografts, are becoming steadily more successful. So far, 14 children have been born worldwide as a result of transplanting frozen/thawed ovarian tissue [13–17]. The very first was born in Belgium in 2004 [13] and the subsequent births have been reported in Israel in 2005 [14], Denmark [15, 16], Spain (birth of twins) [17], and in the USA [18]. One of the patients from Denmark gave birth to two children [19]. Recently, [20] birth from a noncancer patient with thawed ovarian transplants grafted in the pelvic sidewall was reported in France.

Many births have also been reported by using fresh ovarian transplants between monozygotic twins [18]. Ten monozygotic twin pairs requested ovarian transplantation and nine have undergone the procedure (some after failing oocyte donation from their sisters) with cryopreservation of spare tissue. All recipients reinitiated ovulatory menstrual cycles and showed normal day 3 serum FSH levels by 77–142 days. Seven conceived naturally (three twice). Currently, seven healthy babies have been delivered out of ten pregnancies using fresh ovarian tissue transplants. The oldest transplant ceased functioning by 3 years, but this patient conceived again after a second transplant using spare frozen-thawed tissue. Very recently, a birth from the transplant of a whole fresh ovary between two sisters HLA-compatible has also been reported [21].

In summary, when providing an informed consent, it is perfectly legitimate to offer these encouraging but still preliminary results. International fertility preservation society and national special interest groups in both USA and Europe are also closely monitoring the field with follow-up registry.

Conclusions

Cancer survivors may wish to become parents, if they have lost their reproductive function, by using previously stored gametes or gonadal tissue. Fertility preservation serves such a wide range of medico-social circumstances, some quite unique, that patient care requires an individualized and multidisciplinary approach. In particular, fertility specialists offering fertility preservation options to cancer patients should be properly trained and knowledgeable to discuss patient's treatment plan, prognosis, as well as unusual health risks for future offspring and the potential harmful effects of pregnancy.

Overall, there should be no ethical objections to offer these services since they are offered with the scope of preserving future fertility.

However, in practice, there are objections:

1. The options available, except sperm storage and embryo cryopreservation, are all experimental. There is a lack of extended follow-ups about their safety.
2. Posthumous use of stored tissue or gametes. When gametes or tissue is stored for later use, written directives for posthumous use may be given effect, and subsequently born children may be recognized as legal offspring of the deceased. Postmortem reproduction with stored gametes or tissue should be honored when the deceased has given specific consent; programs storing gametes, embryos, or gonadic tissue from cancer patients should inform the options for making advance directives for future use. Whether posthumously conceived or implanted offspring will inherit property from the deceased or will qualify for government benefits will depend on the law of the jurisdiction in which death occurs [22].
3. Concerns about the welfare of offspring resulting from an expected shortened life span of the parent. This concern, however, should not be a sufficient reason to deny cancer survivors assistance in reproducing. Although the effect of the early loss of a parent on a child is regrettable, many children experience stress and sorrow from other circumstances of their birth. The risk that a cancer survivor will die sooner than other parents does not impose an appreciably different burden than the other causes of suffering and unhappiness that persons face in their lives. Protecting such children by preventing their birth altogether is not a reasonable ground for denying cancer survivors' chance to reproduce [22].
4. Concerns about the welfare of children born using gametes frozen after chemotherapy already started.
5. Reseeding of cancer after transplanting cryopreserved tissue.

Future successful production of germ cells de novo could have applications in fertility preservation. Sterile gonads would no longer limit reproduction as it will be possible to produce artificial gametes by dedifferentiation of somatic cells.

References

1. Gosden RG. Fertility preservation-definition, history and prospect. Semin Reprod Med. 2009; 27(6):433–7.
2. SEER cancer statistics review, 1975–2000. National Cancer Institute, Bethesda. http://seer. cancer.gov
3. Bromer JG, Patrizio P. Gynecologic management of fertility. In: Wingard JR, Gastineau D, Leather H, Snyder EL, Szczepiorkowski ZM, editors. Hematopoietic stem cell transplantation: a clinician's handbook. Bethesda: AABB Press; 2009.
4. Patrizio P, Butt S, Caplan A. Ovarian tissue preservation and future fertility: emerging technologies and ethical considerations. J Natl Cancer Inst Monogr. 2005;34:107–10.

5. Hirtz DG, Fitzsimmons LG. Regulatory and ethical issues in the conduct of clinical research involving children. Curr Opin Pediatr. 2002;14:669–75.
6. Grundy R, Larcher V, Gosden RG, Hewitt M, Leiper A, Spoudeas HA, et al. Fertility preservation for children treated for cancer (2): ethics of consent for gamete storage and experimentation. Arch Dis Child. 2001;84:360–2.
7. Grundy R, Larcher V, Gosden RG, Hewitt M, Leiper A, Spoudeas HA, et al. Fertility preservation for children treated for cancer (1): scientific advances and research dilemmas. Arch Dis Child. 2001;84:355–9.
8. Bahadur G. Ethics of testicular stem cells medicine. Hum Reprod. 2004;19:2702–10.
9. Chian RC, Huang JY, Tan SL, et al. Obstetric and perinatal outcome in 200 infants conceived from vitrified oocytes. Reprod Biomed Online. 2008;16:608–10.
10. Noyes N, Porcu E, Boiini A. Over 900 oocyte cryopreservation babies born with no apparent increase in congenital anomalies. Reprod Biomed Online. 2009;18(6):769–76.
11. Sutcliffe AG, Dsouza SW, Cadman J, et al. Minor congenital anomalies, major congenital malformations and development in children conceived from cryopreserved embryos. Hum Reprod. 1995;10:3332–7.
12. Wennerholm UB, Soderstrom-Anttila V, Bergh C, Aittomaki K, Hazekamp J, Nygren KG, et al. Children born after cryopreservation of embryos or oocytes: a systematic review of outcome data. Hum Reprod. 2009;24(9):2158–72.
13. Donnez J, Dolmans MM, Demylle D, et al. Livebirth after orthotopic transplantation of cryopreserved ovarian tissue. Lancet. 2004;364(9443):1405–10.
14. Meirow D, Levron J, Eldar-Geva T, et al. Pregnancy after transplantation of cryopreserved ovarian tissue in a patient with ovarian failure after chemotherapy. N Engl J Med. 2005;353(3):318–21.
15. Demeestere I, Simon P, Emiliani S, Delbaere A, Englert Y. Fertility preservation: successful transplantation of cryopreserved ovarian tissue in a young patient previously treated for Hodgkin's disease. Oncologist. 2007;12(12):1437–42.
16. Andersen CY, Rosendahl M, Byskov AG, et al. Two successful pregnancies following autotransplantation of frozen/thawed ovarian tissue. Hum Reprod. 2008;23:2266–72.
17. Sanchez-Serrano M, Crespo J, Mirabet V, et al. Twins born after transplantation of ovarian cortical tissue and oocyte vitrification. Fertil Steril. 2010;93:268. e11–e13.
18. Silber SJ, DeRosa M, Pineda J, Lenahan K, Grenia D, Gorman K, et al. A series of monozygotic twins discordant for ovarian failure: ovary transplantation (cortical versus microvascular) and cryopreservation. Hum Reprod. 2008;23:1531–7.
19. Ernst E, Bergholdt S, Jorgensen JS, Andersen CY. The first woman to give birth to two children following transplantation of frozen/thawed ovarian tissue. Hum Reprod. 2010;25:1280–1.
20. Roux C, Amiot C, Agnani G, Aubard Y, Rohrlich PS, Piver P. Live birth after ovarian tissue autograft in a patient with sickle cell disease treated by allogeneic bone marrow transplantation. Fertil Steril. 2010;93:2413. e15–e19.
21. Silber SJ, Grudzinskas G, Gosden RG. Successful pregnancy after microsurgical transplantation of an intact ovary. N Engl J Med. 2008;359:2617–8.
22. Robertson JA. Procreative liberty, harm to offspring, and assisted reproduction. Am J Law Med. 2004;30:7–40.

Chapter 3
Spermatogenesis and Testicular Function

Ciler Celik-Ozenci

Testicular germ cell development consists of three fundamental periods: (1) the primordial germ cells (PGCs), the first cells of the germ line lineage in the embryo, (2) spermatogonial stem cells (SSCs), (3) and the spermatozoon, respectively. The first two periods cover the fetal and neonatal periods leading to the formation of the SSCs and include two main types of germ cells, PGCs and gonocytes. PGC differentiates into gonocytes which form type A spermatogonia. The third period is the spermatogenic cycle, a highly regulated chain of events, including mitosis and meiosis, and a differentiation process, such as spermiogenesis, starting with the formation of differentiating spermatogonia and ending with the production of spermatozoa [1]. Accordingly, the course of this chapter, testicular differentiation and spermatogenesis, will be clarified by the following three phases: fetal, neonatal-prepubertal, and postpubertal stages of testis development.

Bipotential Gonads

The primordium of the gonads, during the early stages of fetal life, is undifferentiated; later on sex-determining genes switch on or switch off the expression of genes that contribute to gonadal differentiations which induce the development of the undifferentiated gonads into a testis or an ovary [2]. Induction and patterning of the testis occurs over a brief window of time.

C. Celik-Ozenci, DDS, Ph.D.(✉)
Department of Histology and Embryology, School of Medicine,
Akdeniz University, Campus, Antalya, 07070, Turkey
e-mail: cilerozenci@akdeniz.edu.tr

E. Seli and A. Agarwal(eds.), *Fertility Preservation in Males: Emerging Technologies and Clinical Applications*, DOI 10.1007/978-1-4614-5620-9_3,
© Springer Science+Business Media New York 2012

31

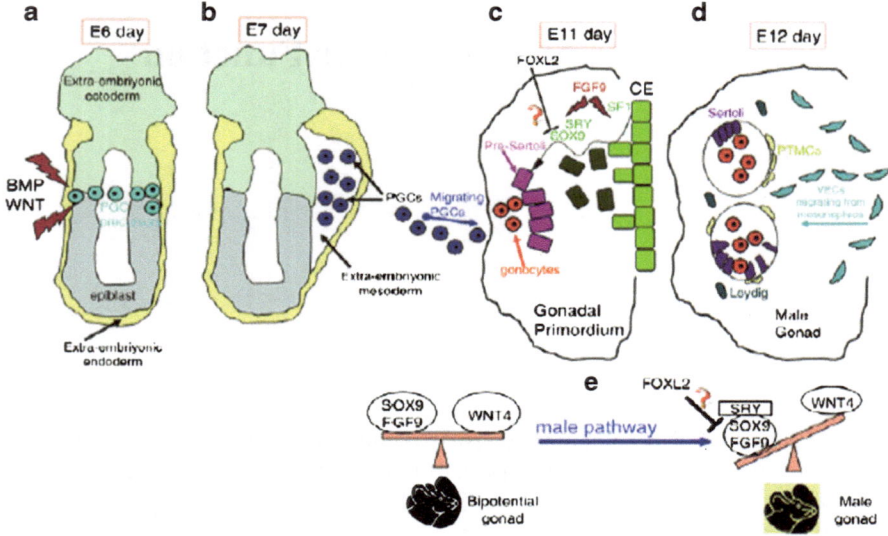

Fig. 3.1 Testicular differentiation during fetal period. (**a**) Bipotential stages cover PGC precursors derived from proximal epiblast cells under the influence of factors, such as BMP and WNT. (**b**) These precursor cells migrate to the extra-embryonic mesoderm, where they are called as PGCs. They continue their migration through the hindgut and dorsal mesentery and reach at the gonadal primordium. (**c**) The migrated PGCs in the gonads now are called gonocytes. Under some factors, such as SF1 and FGF9, coelomic epithelial cells that are indicated as *light green squares*, proliferate and migrate near the gonocytes. They start Sry and Sox9 expression, differentiate into pre-Sertoli cells and start surrounding the gonocytes. A recent question, whether inhibition Foxl2 would contribute to male-specific gene expression (Sry and Sox9) thus regulates Sertoli cell differentiation, needs further investigation. (**d**) Pre-Sertoli cells in the seminiferous cords start their proliferation again and then are called as Sertoli cells. These cells express AMH thus are immature and stop their proliferation and AMH expression while entering puberty. PTMCs surround the seminiferous cords and fetal Leydig cells are seen in the interstitial area between the cords. Vascular endothelial cells migrating from mesonephros contribute to seminiferous cord formation. (**e**) Tight control of the balance between male- and female-specific factors regulates bipotential gonad differentiation into testis. *PGC* primordial germ cell, *CE* coelomic epithelium, *PTMCs* peritubular myoid cells, *VECs* vascular endothelial cells

The origins of the gonads form as genital ridges that comprise thickened coelomic epithelium and are invaded by PGCs [3]. In mice, PGCs are first observed at days 7–8 of pregnancy, in the extra-embryonic mesoderm (Fig. 3.1). The specification of PGCs from pluripotent cells of the embryonic epiblast is dependent on bone morphogenetic protein (BMP) and WNT3 signaling molecules [4–6]. While undergoing repeated divisions, PGCs migrate through the hindgut and dorsal mesentery and arrive at the gonadal primordium on days 10–11 of pregnancy [3]. In humans, by the fourth week of gestation, the PGCs proliferate and migrate from the endoderm of the yolk sac into the undifferentiated gonads, which become morphologically distinct during the seventh week of gestation [7]. Now, the migrated PGCs in the gonads are called gonocytes.

Proliferating coelomic epithelial cells give rise to somatic cells of the differentiating testis which start to proliferate and migrate into the gonads (Fig. 3.1). They fill the gonads and then surround the germ cells, which leads to the formation of sex cords. When the somatic cell proliferation is inhibited, testis cord formation, which is important for the Sertoli cell maintenance, is altered [8]. In the gonad primordium, Sertoli cell precursors form the seminiferous cords at days 12–13 of pregnancy in mice. Within the newly developed cords, Sertoli cells start dividing again. At this stage, it becomes clear whether the gonad becomes a testis or ovary. In the testis, gonocytes proliferate until days 13–14 of pregnancy and then arrest their cell cycles. Their proliferation does not start again until a few days after birth.

Vertebrate gonads are composed of cortex and medulla. The distribution of germ cells in the sex cords is gender-specific [2]. During gonad differentiation into the ovary, germ cells stay in the peripheral region. Thus, the cortex develops better than the medulla and the sex cords disappear. During gonad differentiation into the testis, however, germ cells migrate with the developing sex cords into the gonads, and thus the medulla prevails over the cortex, the rudiment of which transforms into the tunica albuginea.

The organization of the testis structure is also dependent on the migration of mesonephric cells into the gonads (Fig. 3.1). Recent findings indicate that the great majority of cells immigrating into the gonads are endothelial cells [9], but not peritubular myoid cells (PTMCs) as suggested earlier [10–13]. Notably, the testis cord formation occurs at the same time as the development of the testis-specific vascular structure, the coelomic vessel and its branches [9, 14, 15]. Additionally, the onset of steroidogenesis is also critical for the developing testis.

In the ovary, cell proliferation, mesonephric cell migration, testis cord formation, and testis-specific vascular development, as well as steroidogenesis, are inhibited. Primitive sex cords in the ovary disappear and transform into cysts, in which germ cells enter meiosis [2].

Proliferation of Somatic and Germ Cells

Testis determination in most mammals is regulated by a genetic hierarchy initiated by the sex determining region Y (*Sry*) gene. Sry expression is well described in mice sex determination. About 10 day post coitus (dpc), gonads are bipotential but, at around 10.5 dpc and under the effect of the *Sry* gene, the XY gonad primordia start developing as testes and, in the absence of *Sry*, the XX gonad primordia begin to develop as ovaries [16] (Fig. 3.1). Proliferation, which expands the population of Sry-positive cells in the XY gonad, starts at 11.5 dpc and is evident by 12 dpc [8, 17]. *Sry* expression reaches a peak at 11.5 dpc and terminates before 12.5 dpc [18–20]. In the absence of proper expression of *Sry*, presupporting cells develop as granulosa cells [16]. Owing to the proliferation of somatic cells, the testis size becomes twice the size of the ovary after the sex-determining period [2].

Along with the upregulation of Sry-box 9 (*Sox9*) expression, Sry-positive cells turn into Sertoli cells [20]. In mice, *Sry* works together with steroidogenic factor 1 (*SF1*) and activates *Sox9* expression, thus the bipotential gonads differentiate into testis [21] (Fig. 3.1). Sex determination occurs in the human gonad at around 6 weeks of gestation with the development of the testis driven by expression of *Sry* [22]. Different from rodents, *Sry* expression is still visible at the postnatal stage in humans and pigs, particularly in humans; *Sry* is also expressed in germ cells as well as Sertoli cells [23, 24].

Fibroblast growth factors (FGFs) are expressed throughout embryonic development and in a more limited pattern postnatally [25]. FGF signaling has potent effects on cellular differentiation, migration, and morphogenesis, particularly *Fgf9* is necessary for Sertoli cell commitment (Fig. 3.1). *Fgf9* is expressed in both female and male gonads at 11.5 dpc, but *Fgf9* expression has disappeared in the female and enhanced in the male gonad shortly after *Sry* is expressed [26]. Fgf9 promotes the earliest phase of proliferation in coelomic epithelial SF1-positive cells via surface receptor Fgf receptor 2 (Fgfr2) [17, 27]. Fgfr2 is initially found in the cell membrane of coelomic epithelial cells. Fgfr2 seems to mediate proliferation at the surface of the male gonad. On the other hand, it is localized in the nuclei of cells dispersed in the interior of the male gonad but not in the female gonad, thus this receptor in nuclei can induce Sertoli cell differentiation [15, 28].

Recent studies challenge the view that early ovarian development is a default pathway switched on passively by the absence of *Sry* [29]. Interestingly, it has been shown that the mouse ovary continuously represses male-specific genes, from embryonic stages to adulthood [30]. Remarkably, the absence of the key ovarian transcription factor Forkhead box protein L2 (*Foxl2*) induces the activation of repressed male-specific genes during development, postnatally and even during adulthood (Fig. 3.1). Ablation of *Foxl2* in the adult transdifferentiates the supporting cells of the ovary to characteristics of the supporting cells of the testis.

As soon as gonocytes reach the developing male gonad on 10–11 dpc, they continue to proliferate for a short period of time and arrest in G0/G1 phase of the cell cycle at about 12.5–13.5 dpc until after birth [31, 32]. In humans, gonocyte proliferation occurs between the third and sixth months of gestation, followed by a quiescent phase until 2–3 months after birth [33]. Timelines of fetal gonocyte development in mouse, rat, and human is schematized in Fig. 3.2. Eventually, germ line sex determination comes down to the preference between entering meiosis and thus going through to female determination, vs. avoiding meiosis, entering mitotic arrest and progressing to the male determination [34].

It is apparent that meiosis is initiated by retinoic acid (RA), the active derivative of vitamin A, and is avoided at least largely because of the RA-degrading action of a P450 enzyme, CYP26B1 [35–39]. CYP26B1 is initially expressed in developing mice gonads of both sexes, but is then greatly upregulated in the testis and downregulated in the ovary, and thus has been determined as a male-specific meiosis preventing factor [36, 37, 40]. In addition to the degradation of RA in the male fetal gonad, nonretinoid-secreted factors inhibiting both meiosis and mitosis and produced by the testicular somatic cells during fetal and neonatal life have

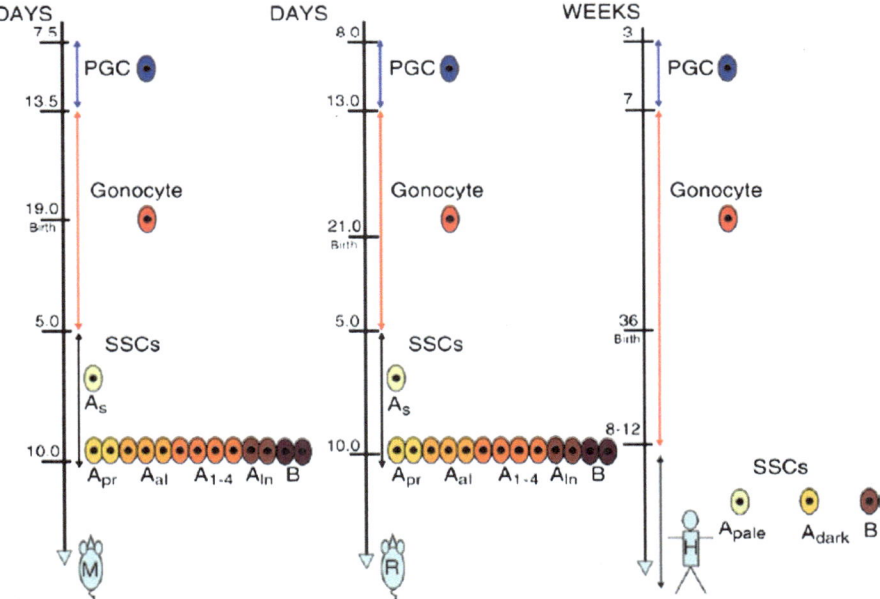

Fig. 3.2 Comparison of the time-lines of fetal and neonatal germ cell development in mouse, rat, and human. *PGCs* primordial germ cells, *SSCs* spermatogonial stem cells, A_s A_{single} spermatogonia, A_{pr} A_{paired} spermatogonia, A_{al} $A_{aligned}$ spermatogonia, A_{1-4} A_{1-4} spermatogonia, A_{In} $A_{intermediate}$ spermatogonia, *B* type B spermatogonia

been found [41]. Using a coculture model in which an undifferentiated female gonad is cultured with a fetal or neonatal testis, it has been demonstrated that the testis prevented the initiation of meiosis and induced male germ cell differentiation in the female gonad.

The source of RA is the mesonephroi to which the gonads are attached [36]. In mice gonads, RA seems to act by inducing the expression of stimulated by RA gene 8 (*Stra8*) [36, 37], a gene considered to be a premeiotic marker [40, 42, 43]. Although a role for RA in initiating meiosis in humans has not been demonstrated yet, meiosis initiates from gonad-mesonephros connection in many mammalian species, including humans [44, 45].

Formation of Sex Cords

The formation of testis cords, which is essential for the maintenance of male testis development, takes place between 11.5 and 12.5 dpc in mice [2]. The creation of Sertoli cell aggregations by their proliferation is related to the expression of extracellular matrix (ECM) and membrane proteins. At this time, the synthesis of cytokeratins begins in Sertoli cells, while desmin is downregulated and becomes synthesized in PTMCs that surround the testis cords [46, 47]. An interaction between the peritubular and Sertoli cells is necessary for basal lamina formation and thus for maintaining the testis cords. It has recently been shown that PTMCs are induced within the gonads and differentiate from interstitial cells derived from the coelomic epithelium [48].

Proteases and their inhibitors, such as testatin, vanin-1, protease nexin-1 (Pn-1) and matricellular protein secreted protein, acidic and rich in cysteine (SPARC), may play key roles in gonadal differentiation. Testatin is specifically expressed in differentiating Sertoli cells in mice. The expression of vanin-1 and Pn-1 is upregulated during sex determination in male gonads [49]. Vanin-1 can regulate the cell adhesion process, which is essential for proper migration and testis cord formation. Pn-1 may be responsible for the maintenance of the integrity of the basal lamina. SPARC is malespecifically expressed during testis development [50] and is localized intracellularly in Sertoli cells, germ cells, and in Leydig cells. SPARC binds to growth factors, platelet-derived growth factors (PDGFs) and FGFs, thereby blocks their action, inhibits cell division and differentiation and alters testis cord and coelomic vessel formation [51, 52].

Testis-specific vascular formation occurs at the same time with testis cord formation [14, 15, 53]. Individual endothelial cells, releasing from the mesonephric vascular plexus located near the gonad–mesonephros border, migrate into the developing testis. They form the wall of the coelomic vessel under the coelomic epithelium, along the cranio-caudal axis of the male gonad. The inhibition of the development of coelomic vessel branches results in the absence of testis cords, which shows that vascular structure is required for the initiation of cord division and also for maintaining the spaces between testis cords [9].

Migration of Mesonephric Cells

Inward migration of mesonephric cells is required for proper cord formation and for maintaining the Sertoli cell differentiation, essential for the growth of the male gonad [54]. The blockage of migration before 11.5 dpc in mice results in the disorganization of testis cords, as well as the coelomic vessel, whereas after 12.5 dpc mesonephric cell migration is not essential for further testis development [13].

PDGFs are diffusible extracellular ligands and they support mesonephric cell migration via their receptors (PDGFRα and PDGFRβ) both of which are expressed in the male gonad. PDGFRα is expressed in gonadal cells throughout the interstitium, and PDGFR is associated with the vasculature [55]. Mutation of *PDGFRa* disrupts signaling between cords and interstitium, thus causes disrupted morphogenesis and lack of fetal Leydig cells (FLCs) [56]. It seems that one of the phenotypes is a secondary result of the other.

Vascular endothelial growth factor (VEGF) may be another key factor in testis development, since blocking its action prevents endothelial cell migration in the testis, as well as coelomic vessel formation and testis cord development [57]. In addition to VEGF, neurotropin (NT) signaling appears to be another required growth factor for proper migration mechanism [58]. Although all of the NT family members and most of their receptors were detected in the embryonic male gonad, neurotropin 3 (NT3) and its receptor neurotrophic tyrosine kinase receptor type (TrkC) were detected in the majority [59].

Steroidogenesis in Embryonic Testis

Sex steroids are crucial regulators of sexual differentiation in males dissimilar to females [60]. Since Leydig cells are the primary major producers of these steroid hormones, maintenance of the normal functions of these cells determines normal development of the testis. FLCs appear in the embryonic gonad shortly after testis determination, and they are present in the mouse testis from 12.5 dpc until they disappear after birth [2]. These cells probably arise from multiple embryonic tissues, including the coelomic epithelium, gonadal ridge mesenchyme, migrating mesonephric cells [61], and probably from fetal adrenal cells [62]. In the human embryonic testis, the precursors of FLCs become functionally active as early as after 6–7 weeks of gestation, at which time testosterone can be detected [63]. The Leydig cells proliferate and differentiate progressively and constantly until they are mature before 19 week of gestation followed by regression [64].

Leydig cell differentiation requires two signaling molecules: PDGF secreted by Sertoli cells and acting via PDGFRα, and Sertoli cells secreted Desert Hedgehog (Dhh) morphogen acting via the Patched 1 (Ptch1) receptor [56, 65]. Although most data concerning the role of Dhh in the regulation of FLCs has been obtained in rodents, there are also reports on the significant role of Dhh signaling in the development of normal testicular phenotype in humans [66, 67]. Besides, PDGF pathway may be involved in the control of Leydig cell development in humans, similar to what has been observed in rodents [68]. Moreover, two additional factors, the transcription factor GATA-4 and the ligands of the insulin-like growth factor (IGF) system, may regulate FLC differentiation both in humans and rodents [69–72]. All of these factors seem to work together to control the differentiation, maturation, and regression of human FLCs.

From Birth to Puberty

From Neonatal Gonocyte to Spermatogonia: Gonocyte Proliferation, Differentiation, and Migration

Testicular gonocytes are the precursors of the SSCs and they correspond to the stage between the PGC and the SSC. However, it is worth mentioning that recent studies have shown that not all but only a fraction of the gonocytes directly differentiates into differentiating spermatogonia [73]. Quiescent fetal gonocytes resume proliferation after birth, and some migrate to the peripheral basement membrane of the seminiferous tubules, where they are termed as spermatogonia.

Neonatal gonocyte proliferation occurs between day post partum (dpp)1 and 4 in the mouse and dpp3 to 4 in the rat [33]. Proliferation in fetal and neonatal gonocytes seems to be differently regulated [74]. From the few studies that have questioned gonocyte proliferation, several factors have emerged as important, including PDGF-BB, 17β-estradiol (E2), leukemia inhibitory factor (LIF), and RA [33]. PDGF and E2 have been shown to be produced by Sertoli cells and are both required to activate gonocyte proliferation at dpp3 [75]. After 3 days in coculture with Sertoli cells, LIF increases the proportion of proliferating rat gonocytes of dpp1, whereas this effect is not seen when gonocytes of dpp3 are cultured [76]. RA is capable of stimulating neonatal rat gonocyte proliferation at dpp3 [77]. Moreover, RA can induce the number of proliferating human fetal gonocytes from 6- to 10-weekold fetuses after 4 days in organ culture [78].

Differences between a neonatal gonocyte and a spermatogonium are their morphological appearances and their different locations within the seminiferous cord, gonocytes being large spherical cells at the center of the cords while spermatogonia are smaller half-moon shaped cells positioned along the basement membrane [33]. In fact, location by itself is not a sufficient criterion and many of the proteins expressed in gonocytes are also present in spermatogonia. Recent advances in the characterization of SSCs have revealed new gene markers that might help discrimination between the two stages.

One of these genes Stra8, an indicator of gonocyte to spermatogonia transition, is expressed in spermatogonia and premeiotic cells, but not in postmeiotic or Sertoli cells [79]. A progressive increase of Stra8 in the germ cells of the first spermatogonial wave, starting with a few positive spermatogonia at dpp5, followed by an increasing number of positive preleptotenes and early leptotene spermatocytes over time has been shown in postnatal testis. Recently, RA has been identified as a key regulator of gonocyte differentiation in rats [80] and in mice [81]. RA makes this action by reducing the proportion of SSCs via decreasing growth factor receptor alpha 1 (GFRα1) transcript and by increasing the proportion of type A spermatogonia via increasing c-kit mRNA in these cells [80]. It may also be presumed that the same mechanisms might have a role in human gonocyte differentiation.

In order to differentiate, gonocytes should migrate from the center of the seminiferous cords to the basement membrane [82]. c-kit is transiently expressed in subsets of gonocytes presenting pseudopods and seems to regulate gonocyte migration toward the base of the cords [83]. After SSCs locate at the basement membrane of the cords c-kit is downregulated in these nonmigrating cells. ADAM-integrin-tetraspanin complexes, which are expressed in gonocytes, have been proposed as potential mediators of gonocyte migration [82]. A disintegrin and a metalloprotease domain (ADAM) 1 and ADAM 2 appear at dpp1 (ADAM1) and dpp2 (ADAM2) mice gonocytes [84]. Recently, Basciani et al. observed that a subset of the gonocytes migrated to the basement membrane regardless of the PDGFR and c-kit inhibitor, imatinib [85]. Migrated gonocytes could proliferate and differentiate, thus they could correspond to a pool of gonocytes that will differentiate into SSCs. Thus, at least two populations of neonatal gonocytes could coexist in the newborn testis [86–88]. First, gonocytes which will become SSCs and will populate the SSC niche. Second, a group of gonocytes that will differentiate to A pair stage of spermatogonia.

SSCs: Self Renewal, Differentiation, and SSC Niche

Self-renewal and differentiation of SSCs, a subpopulation of type A spermatogonia laying on the basement membrane of seminiferous tubules, are key events for normal spermatogenesis. A better understanding of molecular mechanisms in this population may help to distinguish factors which contribute to male infertility and testicular cancer.

There are several subtypes of A spermatogonia in rodents, including the A_{single} (A_s), the A_{paired} (A_{pr}), the $A_{aligned}$ (A_{al}), and the A_{1-4} spermatogonia based on differences in their morphology and phenotype [89, 90]. A_{single} spermatogonia act as stem cells, whereas interconnected spermatogonia are committed to differentiation and lose their stem cell potential. In mice, after mitotic divisions, cell cysts are formed by intercellular bridges between the cells (A_{al} spermatogonia) and they most probably enter meiosis synchronously. The A_{al} spermatogonia, in turn, give rise to several generations of spermatogonia, including type A_{1-4}, intermediate, and type B spermatogonia [89] (Fig. 3.2). This model, established on the basis of the morphological observation of fixed specimens, has provided the theoretical basis for germ line stem cell biology, regardless of the animal species. However, a logical shortcoming of this model is that it has not been derived from the behavior of cells. Type B cells generate spermatocytes, spermatids, and mature sperm.

In humans, A_{dark} and A_{pale} spermatogonia are distinguished by Clermont [91–94] (Fig. 3.2). It has been considered that A_{dark} cells are reserve stem cells, while the A_{pale} cells are renewing stem cells, a concept which is also adopted for monkey spermatogonia. An adult rhesus testis contains roughly equal numbers of A_{dark} and A_{pale} spermatogonia [95], where A_{dark} appear to fulfill the role of a "reserve stem cell." However, the functional identity of the human SSCs is still unknown. For extensive reading about key signaling molecules and their physiological roles in regulating cell fate decisions of spermatogonia, you may read the review by He et al. [96].

Intrinsic and extrinsic factors tightly control self-renewal and differentiation of SSCs (Fig. 3.3). Among the intrinsic regulators is promyelocytic leukemia zinc finger (Plzf), a transcriptional repressor expressed in the $A_s/A_{pr}/A_{al}$ population. In mice lacking Plzf, spermatogenesis is initially established but is then gradually lost within months, indicating that Plzf is crucial for the maintenance of stem cell activity rather than stem cell founding [97, 98]. Recently, it has been discovered that another intrinsic factor, Nanos2, plays a central role in the maintenance of self-renewal in mice SSCs [99, 100]. The researchers showed that Nanos2 is expressed

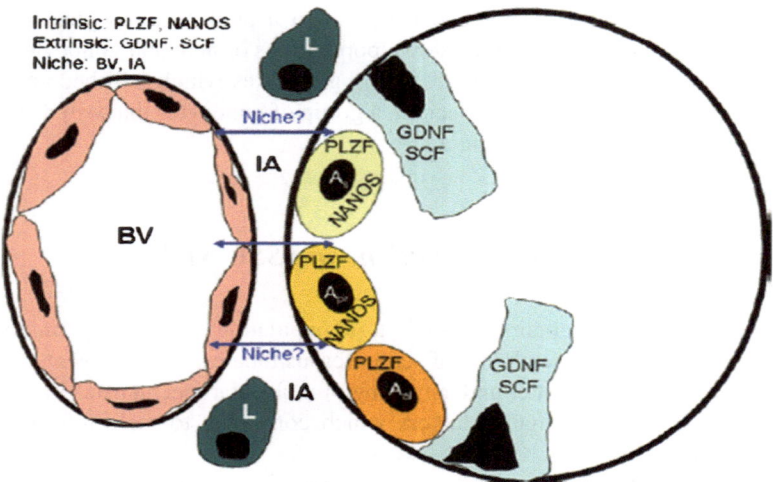

Fig. 3.3 Factors that control self-renewal and differentiation of SSCs. Some important intrinsic factors derived from spermatogonia, and extrinsic factors secreted from Sertoli cells are known to regulate the stem cell pool and differentiation. Since the testicular stem cell population at the base of the seminiferous tubules is localized close to blood vessels and interstitial area, it is possible to argue that this niche might contribute to their proliferation and differentiation processes as well. *BV* blood vessel, *L* Leydig cell, *IA* interstitial area

in a majority of A_s and A_{pr} spermatogonia and its constitutive expression inhibits spermatogenic differentiation. On the other hand, depletion of Nanos2 causes a loss of stem cell maintenance.

In the animal body, self-renewal and differentiation of stem cells are under the control of a specialized microenvironment called the stem cell niche [101–103]. Existing knowledge suggests that the $A_s/A_{pr}/A_{al}$ population is localized at the vicinity of the blood vessels and interstitium around the seminiferous tubules and that they leave this niche upon differentiation into A_1 spermatogonia and migrate to the lumen of the tubule [87, 104, 105]. Although it needs further investigation; this region could be specialized to provide a microenvironmental niche for SSCs. On the other hand, factors, which have been secreted extracellularly, play essential roles in the stem cell-niche interactions.

Glial cell line-derived neurotrophic factor (GDNF), secreted by Sertoli cells, regulates the self-renewal and differentiation of mice SSCs in a dose-dependent manner. In vivo, GDNF overexpression causes the accumulation of undifferentiated spermatogonia, and on the contrary, GDNF ablation results in the reduction of spermatogonia numbers [106], suggesting that GDNF is vital for the self-renewal and maintenance of SSCs. GDNF acts via a receptor complex containing GDNF family receptor alpha-1 (GFRA1) and rearranged during transfection (RET), both of which are expressed in SSCs, Apr cells [107–109], and A_{al} spermatogonia [110]. Loss of GDNF/GFRA1 signaling triggers differentiation of SSCs into A_{1-4} spermatogonia [109].

Stem cell factor (SCF, KIT ligand, steel factor) may play a diverse role in regulating the fate of mice type A_{1-4} spermatogonia [111]. After birth, there was a striking elevation in SCF transcript in whole mice testes at day 5, corresponding to the time when proliferation of Type A spermatogonia is initiated. SCF protein was present within Sertoli cells at 1–7 dpp and in the adult [112] suggesting that a significant change in Sertoli cell function occurs at this time, which is probably linked to the transition of gonocytes to spermatogonia and expansion of the latter population by mitosis.

SSC-specific marker has not been identified for any species so far [113]. However, the combined expression profiles of multiple markers provide information about stem, progenitor, and differentiating spermatogonia in rodents that may be useful for identifying primate SSCs. Expression of rodent spermatogonial markers (GFRA1, PLZF, NGN3, and KIT) were recently identified in the adult rhesus testis, and it has been found that most A_{dark} and almost 50% of A_{pale} exhibit the phenotype GFRA1+, PLZF+, NGN3−, and KIT− those could be either stem or A_{al} progenitor type spermatogonia. Based on the conservation of molecular markers from rodents to primates, it has been proposed that the stem cell pool is considerably larger in rhesus than mice testes. In contrast to the large SSC pool, the relative small size of the progenitor pool (GFRA1+, PLZF+, NGN3+, KIT−) has been identified in adult macaques compared to rodents. Thus, it appears that rodents may have few SSCs and more progenitors while rhesus testes may have more SSCs and fewer progenitors.

When comparative analysis of Plzf, SSC maintenance marker, was carried out in mice, rhesus monkeys, and human testes, it was found that Plzf staining in human testes was restricted to some cells on the basement membrane of the seminiferous epithelium similar to mice and monkeys and the number of Plzf cells was more comparable to monkeys than mice. By analyzing expressions of POU5F1, TSPY, and KIT, other differences in marker expression have been reported between rodent and human spermatogonia, suggesting that the dynamics of the SSC pools in humans are similar to monkeys [114].

Initiation of Meiosis, Building the Barrier, and Entering Puberty

Many components of the meiosis are shared between males and females during meiotic divisions; however, as mentioned in earlier sections of this chapter, the timing and regulation of meiosis differ between the sexes [115, 116]. In females, meiotic divisions initiate during a brief window in embryonic development, whereas meiotic initiation in males is first observed at puberty and then reappears repeatedly and continuously throughout adulthood. In male mice, in addition to the mitotically active spermatogonia Stra8 is expressed in preleptotene spermatocytes, the most advanced cell type before meiotic prophase [43, 117]. It has been discovered that similar to the initiation of ovarian germ cell meiosis, Stra8 regulates meiotic initiation through RA induction of Stra8 in juvenile mice testes, phenomenon

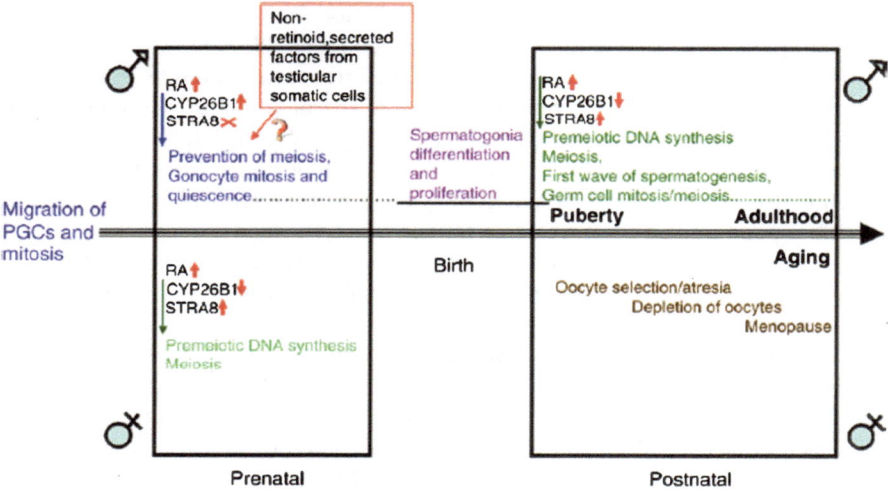

Fig. 3.4 Schematic representation of meiosis regulation during prenatal and postnatal stages in male and female

so prominently shared between two sexes, suggesting that the molecular control of premeiosis S-phase entry is conserved and gender-neutral [39] (Fig. 3.4).

The proliferation of type B spermatogonia results in the production of primary spermatocytes that enter and undergo the first meiotic division [118]. Meiosis is composed of prophase, metaphase, anaphase, and telophase. Prophase consists of the following defined cells: preleptotene spermatocytes, the first cells to be formed; leptotene spermatocytes, where chromosomes become more apparent as long threads; zygotene spermatocytes, where chromosomal pairing and synapsis is initiated; and pachytene spermatocytes, where pairing of chromosomes is completed. Desynapsis occurs in diplotene spermatocytes and chromosomes partially separate. Diakinesis involves shortening of chromosomes with each one composed of two chromatids. The synaptonemal complex is a proteinaceous scaffold that connects homologous chromosomes along their entire length to form synapsed bivalents during meiotic prophase I of spermatocytes. This complex mediates the synapsis of homologous chromosomes. Prophase is followed by metaphase where attachment of paired chromatids to the equator of the spindle takes place. Anaphase involves movement of paired chromatids to opposite poles of the spindle, and telophase gives rise to daughter cells with a haploid number of chromosomes with diploid DNA content called secondary spermatocytes. These cells have a short half life and they go through the second meiotic division to form spermatids with a haploid DNA content and a haploid chromosomal number. The duration of spermatocyte maturation takes around 21 days in rats and 25 days in humans [119].

Spermatogenesis requires coordination of germ cell movement and differentiation which essentially depends on interactions between Sertoli and germ cells. The leptotene spermatocytes must cross the blood-testis barrier (BTB) before the

G2–MI transition of their cell cycle and enter the adluminal compartment of the seminiferous epithelium [120]. The movement of germ cells across the seminiferous epithelium requires distinct kinases and phosphatases. Cyclins and cyclin-dependent kinases, polo-like and aurora kinases, and the mitogen-activated protein kinase (MAPK) signaling pathway regulate the transit of primary spermatocytes across the BTB and contribute to BTB remodeling during germ cell divisions. Additionally, transforming growth factor β (TGF-β) can promote cell cycle arrest thus misregulation of its activity can affect cell division, disturb cell adhesion and increase cell movement. Cell cycle events are likely to participate in the movement of leptotene spermatocytes across the BTB. Cytokines and testosterone also have important roles during this process. Nevertheless, additional proteins and signaling cascades that regulate the entry of differentiating leptotene spermatocytes into the adluminal compartment are yet to be determined.

Puberty and Sperm Production

Spermatogenesis can be divided into premeiotic, meiotic, and postmeiotic phases. The premeiotic phase is characterized by an increase in cell numbers as a result of mitotic divisions of diploid spermatogonia. The meiotic prophase then leads to the formation of haploid round spermatids. The postmeiotic phase (spermiogenesis) involves the morphogenetic events that are necessary for sperm production.

Spermiogenesis is a metamorphosis process involving the maturation and differentiation of the early haploid male gamete to a mature spermatozoon. The spermiogenetic period takes about 2–3 weeks in mice [121] and 5–6 weeks in humans [91, 122]. During spermiogenesis of round spermatids, the nucleus takes on a more compact shape, the mitochondria are rearranged, the flagellum develops and the acrosome is generated [1].

Striking chromatin remodeling and genome reorganization takes place during the postmeiotic maturation of round spermatids. The regulation of gene expression in postmeiotic male germ cells is governed by specific mechanisms unique to these cells. In cAMP-dependent signaling pathway, gene expression is mediated primarily by two molecules: the cAMP-response element binding protein (CREB) and the cAMP-responsive element modulator (CREM) [123]. The transcriptional activator CREM is highly expressed in postmeiotic cells [124–126], and CREM deficiency results in the lack of postmeiotic cellspecific gene expression. CREM mutant male mice show postmeiotic arrest at the first step of spermiogenesis, where late spermatids were completely absent [127]. CREM activity is regulated through interactions with a germ cell-specific, the CREM phosphorylation-independent transcriptional coactivator, the activator of CREM in testis (ACT) [128]. The ability of ACT to regulate CREM activity is controlled by a germ cellspecific kinesin, Kif17b, which regulates the subcellular distribution of ACT. The CREM gene is expressed also in human germ cells and a switch from the expression of repressors to activators is present in normospermic men [129]. Conversely, in patients showing a testicular pattern of round spermatid maturation arrest only, CREM repressors are expressed. Thus, it has been proposed that CREM plays a role also in human spermatogenesis and that the absence of "the CREM switch" can be associated to spermatogenic arrest.

During spermiogenesis, spermatids repackage their DNA with highly basic arginine and cysteine rich protamines, but in almost all species a small residue of histone-bound DNA is kept and gains access to the ooplasm [130]. Despite the fundamental nature of these events, the molecular basis of the driving elements controlling them is not known. The entry of core histones alongside modified histones and histone variants to the egg leaves room for DNA and histone-based epigenetic signals that may be important for subsequent embryonic development. Recent evidence, arising from a closer examination of the composition of histone-bound and protamine bound domains in human and murine spermatozoa, indicates that the former chromatin is indeed the more significant contributor to a novel epigenetic signal in these cells [131–135].

Although the sperm nucleus is transcriptionally quiet, it contains diverse RNA populations, mRNAs, antisense, and miRNAs that have been transcribed throughout spermatogenesis [136]. There is also an endogenous reverse transcriptase that may be activated under certain circumstances. It is now commonly accepted that sperm can deliver some RNAs to the oocyte at fertilization. Some recent data are provided, supporting the view that analyzing the profile of spermatozoal RNAs could be useful for the assessment of male fertility. The importance of paternal RNAs for spermatogenesis and subsequently for male fertility is yet to be determined.

Conclusions

Spermatogenesis is characterized by three specific phases: proliferation and differentiation, meiosis, and morphogenesis of haploid germ cells which takes approximately 1 month in mice and 2 months in humans. Sex determination occurs in the fetal gonads during development. Germ cells enter meiosis, and thus commit to the oogenic pathway, or stay away from meiosis, enter a state of quiescence and commit to the spermatogenic pathway. The main clinical consequence of defects in testicular differentiation during fetal periods is likely to be infertility. Another outcome of this condition in humans is increased susceptibility to testicular germ cell tumors, which have been proposed to originate from impaired or delayed germ cell differentiation during fetal testis development.

Chemotherapy and radiotherapy, even in low doses, may have a detrimental effect on the seminiferous epithelium and disrupt spermatogenesis in both children and adults [137]. Many chemotherapeutic drugs cross the BTB and act by damaging rapidly proliferating cells, therefore exerting their gonadotoxic effect. Exposure to chemotherapy may also result in the induction of sperm DNA fragmentation [138]. The extent of damage to germ cells depends on the class of chemotherapeutic agent, dosage, spermatogenetic stage targeted as well as the original pretreatment fertility potential of the patient. Cytotoxic therapy influences spermatogenesis at least temporarily and in some cases permanently. Recovery of spermatogenesis depends on the drugs used and on the cumulative dose given. Germ cells are also very sensitive to irradiation while the Leydig cells are more resistant owing to their slower rate of turnover [139]. Radiotherapy affects sperm concentration; moreover, irradiation increases sperm DNA damage [140].

In conclusion, the seeds of functional testes are built at prenatal life, progress during prepubertal ages, are upregulated after puberty and continue lifelong in males. Fortunately, mechanisms regulating these events are being discovered by researchers perpetually and much more still remain to be uncovered. The gonadotoxic side effect of cancer therapy is one of the challenges yet to be conquered. Since prepubertal age is not a silent period in testicular development, the effects of chemotherapy and radiotherapy on the prepubertal gonads seem just as detrimental as later in life.

References

1. Russell LDER, Sinha HAP, Clegg ED. Mammalian spermatogenesis. In: Russell LD, Ettlin RA, Sinha HAP, Clegg ED, editors. Histological and histopathological evaluation of the testis. Clearwater: Cache River Press; 1990. p. 1–38.
2. Piprek RP. Molecular and cellular machinery of gonadal differentiation in mammals. Int J Dev Biol. 2010;54(5):779–86.
3. McLaren A. Primordial germ cells in the mouse. Dev Biol. 2003;262(1):1–15.
4. Surani MA, Hayashi K, Hajkova P. Genetic and epigenetic regulators of pluripotency. Cell. 2007;128(4):747–62.
5. Saitou M. Specification of the germ cell lineage in mice. Front Biosci. 2009;14:1068–87.
6. Ohinata Y, Ohta H, Shigeta M, et al. A signaling principle for the specification of the germ cell lineage in mice. Cell. 2009;137(3):571–84.
7. Shalet SM. Normal testicular function and spermatogenesis. Pediatr Blood Cancer. 2009;53(2):285–8.
8. Schmahl J, Capel B. Cell proliferation is necessary for the determination of male fate in the gonad. Dev Biol. 2003;258(2):264–76.
9. Combes AN, Wilhelm D, Davidson T, et al. Endothelial cell migration directs testis cord formation. Dev Biol. 2009;326(1):112–20.
10. Buehr M, Gu S, McLaren A. Mesonephric contribution to testis differentiation in the fetal mouse. Development. 1993;117(1):273–81.
11. Martineau J, Nordqvist K, Tilmann C, et al. Malespecific cell migration into the developing gonad. Curr Biol. 1997;7(12):958–68.
12. Merchant-Larios H, Moreno-Mendoza N, Buehr M. The role of the mesonephros in cell differentiation and morphogenesis of the mouse fetal testis. Int J Dev Biol. 1993;37(3): 407–15.
13. Tilmann C, Capel B. Mesonephric cell migration induces testis cord formation and Sertoli cell differentiation in the mammalian gonad. Development. 1999;126(13):2883–90.
14. Brennan J, Karl J, Capel B. Divergent vascular mechanisms downstream of Sry establish the arterial system in the XY gonad. Dev Biol. 2002;244(2):418–28.
15. Brennan J, Capel B. One tissue, two fates: molecular genetic events that underlie testis versus ovary development. Nat Rev Genet. 2004;5(7):509–21.
16. Koopman P, Gubbay J, Vivian N, et al. Male development of chromosomally female mice transgenic for Sry. Nature. 1991;351(6322):117–21.

17. Schmahl J, Eicher EM, Washburn LL, et al. Sry induces cell proliferation in the mouse gonad. Development. 2000;127(1):65–73.
18. Hacker A, Capel B, Goodfellow P, et al. Expression of Sry, the mouse sex determining gene. Development. 1995;121(6):1603–14.
19. Bullejos M, Koopman P. Spatially dynamic expression of Sry in mouse genital ridges. Dev Dyn. 2001;221(2):201–5.
20. Sekido R, Bar I, Narvaez V, et al. SOX9 is up-regulated by the transient expression of SRY specifically in Sertoli cell precursors. Dev Biol. 2004;274(2):271–9.
21. Sekido R, Lovell-Badge R. Sex determination involves synergistic action of SRY and SF1 on a specific Sox9 enhancer. Nature. 2008;453(7197):930–4.
22. Houmard B, Small C, Yang L, et al. Global gene expression in the human fetal testis and ovary. Biol Reprod. 2009;81(2):438–43.
23. Parma P, Pailhoux E, Cotinot C. Reverse transcription-polymerase chain reaction analysis of genes involved in gonadal differentiation in pigs. Biol Reprod. 1999;61(3):741–8.
24. Salas-Cortes L, Jaubert F, Barbaux S, et al. The human SRY protein is present in fetal and adult Sertoli cells and germ cells. Int J Dev Biol. 1999;43(2):135–40.
25. Itoh N. The Fgf families in humans, mice, and zebrafish: their evolutional processes and roles in development, metabolism, and disease. Biol Pharm Bull. 2007;30(10):1819–25.
26. Kim Y, Kobayashi A, Sekido R, et al. Fgf9 and Wnt4 act as antagonistic signals to regulate mammalian sex determination. PLoS Biol. 2006;4(6):e187.
27. Karl J, Capel B. Sertoli cells of the mouse testis originate from the coelomic epithelium. Dev Biol. 1998;203(2):323–33.
28. Kim Y, Bingham N, Sekido R, et al. Fibroblast growth factor receptor 2 regulates proliferation and Sertoli differentiation during male sex determination. Proc Natl Acad Sci USA. 2007;104(42):16558–63.
29. Veitia RA. FOXL2 versus SOX9: a lifelong "battle of the sexes". Bioessays. 2010;32(5):375–80.
30. Uhlenhaut NH, Jakob S, Anlag K, et al. Somatic sex reprogramming of adult ovaries to testes by FOXL2 ablation. Cell. 2009;139(6):1130–42.
31. Hilscher B, Hilscher W, Bulthoff-Ohnolz B, et al. Kinetics of gametogenesis. I. Comparative histological and autoradiographic studies of oocytes and transitional prospermatogonia during oogenesis and prespermatogenesis. Cell Tissue Res. 1974;154(4):443–70.
32. McLaren A. Meiosis and differentiation of mouse germ cells. Symp Soc Exp Biol. 1984;38:7–23.
33. Culty M. Gonocytes, the forgotten cells of the germ cell lineage. Birth Defects Res C Embryo Today. 2009;87(1):1–26.
34. Bowles J, Koopman P. Sex determination in mammalian germ cells: extrinsic versus intrinsic factors. Reproduction. 2010;139(6):943–58.
35. Baltus AE, Menke DB, Hu YC, et al. In germ cells of mouse embryonic ovaries, the decision to enter meiosis precedes premeiotic DNA replication. Nat Genet. 2006;38(12):1430–4.
36. Bowles J, Knight D, Smith C, et al. Retinoid signaling determines germ cell fate in mice. Science. 2006;312(5773):596–600.
37. Koubova J, Menke DB, Zhou Q, et al. Retinoic acid regulates sex-specific timing of meiotic initiation in mice. Proc Natl Acad Sci USA. 2006;103(8):2474–9.
38. MacLean G, Li H, Metzger D, et al. Apoptotic extinction of germ cells in testes of Cyp26b1 knockout mice. Endocrinology. 2007;148(10):4560–7.
39. Anderson EL, Baltus AE, Roepers-Gajadien HL, et al. Stra8 and its inducer, retinoic acid, regulate meiotic initiation in both spermatogenesis and oogenesis in mice. Proc Natl Acad Sci USA. 2008;105(39):14976–80.
40. Menke DB, Page DC. Sexually dimorphic gene expression in the developing mouse gonad. Gene Expr Patterns. 2002;2(3–4):359–67.
41. Guerquin MJ, Duquenne C, Lahaye JB, et al. New testicular mechanisms involved in the prevention of fetal meiotic initiation in mice. Dev Biol. 2010;346(2):320–30.
42. Bouillet P, Oulad-Abdelghani M, Vicaire S, et al. Efficient cloning of cDNAs of retinoic acid-responsive genes in P19 embryonal carcinoma cells and characterization of a novel mouse gene, Stra1 (mouse LERK-2/Eplg2). Dev Biol. 1995;170(2):420–33.

43. Oulad-Abdelghani M, Bouillet P, Decimo D, et al. Characterization of a premeiotic germ cell-specific cytoplasmic protein encoded by Stra8, a novel retinoic acid-responsive gene. J Cell Biol. 1996;135(2):469–77.
44. Byskov AG. The role of the rete ovarii in meiosis and follicle formation in the cat, mink and ferret. J Reprod Fertil. 1975;45(2):201–9.
45. Bendsen E, Byskov AG, Andersen CY, et al. Number of germ cells and somatic cells in human fetal ovaries during the first weeks after sex differentiation. Hum Reprod. 2006;21(1):30–5.
46. Virtanen I, Kallajoki M, Narvanen O, et al. Peritubular myoid cells of human and rat testis are smooth muscle cells that contain desmin-type intermediate filaments. Anat Rec. 1986;215(1): 10–20.
47. Frojdman K, Paranko J, Virtanen I, et al. Intermediate filaments and epithelial differentiation of male rat embryonic gonad. Differentiation. 1992;50(2):113–23.
48. Cool J, Carmona FD, Szucsik JC, et al. Peritubular myoid cells are not the migrating population required for testis cord formation in the XY gonad. Sex Dev. 2008;2(3):128–33.
49. Grimmond S, Van Hateren N, Siggers P, et al. Sexually dimorphic expression of protease nexin-1 and vanin-1 in the developing mouse gonad prior to overt differentiation suggests a role in mammalian sexual development. Hum Mol Genet. 2000;9(10):1553–60.
50. Wilson MJ, Bowles J, Koopman P. The matricellular protein SPARC is internalized in Sertoli, Leydig, and germ cells during testis differentiation. Mol Reprod Dev. 2006;73(5):531–9.
51. Yan Q, Sage EH. SPARC, a matricellular glycoprotein with important biological functions. J Histochem Cytochem. 1999;47(12):1495–150653.
52. Brekken RA, Sage EH. SPARC, a matricellular protein: at the crossroads of cell-matrix communication. Matrix Biol. 2001;19(8):816–27.
53. Coveney D, Cool J, Oliver T, et al. Four-dimensional analysis of vascularization during primary development of an organ, the gonad. Proc Natl Acad Sci USA. 2008;105(20):7212–7.
54. Capel B, Albrecht KH, Washburn LL, et al. Migration of mesonephric cells into the mammalian gonad depends on Sry. Mech Dev. 1999;84(1–2):127–31.
55. Cool J, Capel B. Mixed signals: development of the testis. Semin Reprod Med. 2009;27(1): 5–13.
56. Brennan J, Tilmann C, Capel B. Pdgfr-alpha mediates testis cord organization and fetal Leydig cell development in the XY gonad. Genes Dev. 2003;17(6):800–10.
57. Cool J. Testis morphogenesis requires VEGF mediated endothelial migration via a novel mechanism of vascular remodelling. In: Fifth international symposium on vertebrate sex determination, Kona; 2009.
58. Cupp AS, Kim GH, Skinner MK. Expression and action of neurotropin-3 and nerve growth factor in embryonic and early postnatal rat testis development. Biol Reprod. 2000;63(6):1617–28.
59. Levine E, Cupp AS, Skinner MK. Role of neurotropins in rat embryonic testis morphogenesis (cord formation). Biol Reprod. 2000;62(1):132–42.
60. Svechnikov K, Landreh L, Weisser J, et al. Origin, development and regulation of human Leydig cells. Horm Res Paediatr. 2010;73(2):93–101.
61. Habert R, Lejeune H, Saez JM. Origin, differentiation and regulation of fetal and adult Leydig cells. Mol Cell Endocrinol. 2001;179(1–2):47–74.
62. O'Shaughnessy PJ, Baker PJ, Johnston H. The foetal Leydig cell – differentiation, function and regulation. Int J Androl. 2006;29(1):90–5. discussion 105–8.
63. Tapanainen J, Kellokumpu-Lehtinen P, Pelliniemi L, et al. Age-related changes in endogenous steroids of human fetal testis during early and midpregnancy. J Clin Endocrinol Metab. 1981;52(1):98–102.
64. Haider SG. Cell biology of Leydig cells in the testis. Int Rev Cytol. 2004;233:181–241.
65. Pierucci-Alves F, Clark AM, Russell LD. A developmental study of the Desert Hedgehog-null mouse testis. Biol Reprod. 2001;65(5):1392–402.
66. Canto P, Vilchis F, Soderlund D, et al. A heterozygous mutation in the Desert Hedgehog gene in patients with mixed gonadal dysgenesis. Mol Hum Reprod. 2005;11(11):833–6.
67. Fowler PA, Cassie S, Rhind SM, et al. Maternal smoking during pregnancy specifically reduces human fetal Desert Hedgehog gene expression during testis development. J Clin Endocrinol Metab. 2008;93(2):619–26.

68. Gnessi L, Basciani S, Mariani S, et al. Leydig cell loss and spermatogenic arrest in platelet-derived growth factor (PDGF)-A-deficient mice. J Cell Biol. 2000;149(5):1019–26.
69. Ketola I, Pentikainen V, Vaskivuo T, et al. Expression of transcription factor GATA-4 during human testicular development and disease. J Clin Endocrinol Metab. 2000;85(10):3925–31.
70. Bielinska M, Seehra A, Toppari J, et al. GATA-4 is required for sex steroidogenic cell development in the fetal mouse. Dev Dyn. 2007;236(1):203–13.
71. Rouiller-Fabre V, Lecref L, Gautier C, et al. Expression and effect of insulin-like growth factor I on rat fetal Leydig cell function and differentiation. Endocrinology. 1998;139(6):2926–34.
72. Berensztein EB, Baquedano MS, Pepe CM, et al. Role of IGFs and insulin in the human testis during post natal activation: differentiation of steroidogenic cells. Pediatr Res. 2008;63(6):662–5.
73. Yoshida S, Sukeno M, Nakagawa T, et al. The first round of mouse spermatogenesis is a distinctive program that lacks the self-renewing spermatogonia stage. Development. 2006;133(8):1495–505.
74. Ohmura M, Naka K, Hoshii T, et al. Identification of stem cells during prepubertal spermatogenesis via monitoring of nucleostemin promoter activity. Stem Cells. 2008;26(12): 3237–46.
75. Thuillier R, Wang Y, Culty M. Prenatal exposure to estrogenic compounds alters the expression pattern of platelet-derived growth factor receptors alpha and beta in neonatal rat testis: identification of gonocytes as targets of estrogen exposure. Biol Reprod. 2003;68(3): 867–80.
76. De Miguel MP, De Boer-Brouwer M, Paniagua R, et al. Leukemia inhibitory factor and ciliary neurotropic factor promote the survival of Sertoli cells and gonocytes in coculture system. Endocrinology. 1996;137(5):1885–93.
77. Livera G, Rouiller-Fabre V, Durand P, et al. Multiple effects of retinoids on the development of Sertoli, germ, and Leydig cells of fetal and neonatal rat testis in culture. Biol Reprod. 2000;62(5):1303–14.
78. Lambrot R, Coffigny H, Pairault C, et al. Use of organ culture to study the human fetal testis development: effect of retinoic acid. J Clin Endocrinol Metab. 2006;91(7):2696–703.
79. Giuili G, Tomljenovic A, Labrecque N, et al. Murine spermatogonial stem cells: targeted transgene expression and purification in an active state. EMBO Rep. 2002;3(8):753–9.
80. Wang Y, Culty M. Identification and distribution of a novel platelet-derived growth factor receptor beta variant: effect of retinoic acid and involvement in cell differentiation. Endocrinology. 2007;148(5):2233–50.
81. Zhou Q, Li Y, Nie R, et al. Expression of stimulated by retinoic acid gene 8 (Stra8) and maturation of murine gonocytes and spermatogonia induced by retinoic acid in vitro. Biol Reprod. 2008;78(3):537–45.
82. Tres LL, Kierszenbaum AL. The ADAM-integrin-tetraspanin complex in fetal and postnatal testicular cords. Birth Defects Res C Embryo Today. 2005;75(2):130–41.
83. Orth JM, Qiu J, Jester WF, et al. Expression of the c-kit gene is critical for migration of neonatal rat gonocytes in vitro. Biol Reprod. 1997;57(3):676–83.
84. Rosselot C, Kierszenbaum AL, Rivkin E, et al. Chronological gene expression of ADAMs during testicular development: prespermatogonia (gonocytes) express fertilin beta (ADAM2). Dev Dyn. 2003;227(3):458–67.
85. Basciani S, De Luca G, Dolci S, et al. Platelet-derived growth factor receptor beta-subtype regulates proliferation and migration of gonocytes. Endocrinology. 2008;149(12):6226–35.
86. Shinohara T, Orwig KE, Avarbock MR, et al. Germ line stem cell competition in postnatal mouse testes. Biol Reprod. 2002;66(5):1491–7.
87. Yoshida S, Sukeno M, Nabeshima Y. A vasculature associated niche for undifferentiated spermatogonia in the mouse testis. Science. 2007;317(5845):1722–6.
88. Seandel M, Falciatori I, Shmelkov SV, et al. Niche players: spermatogonial progenitors marked by GPR125. Cell Cycle. 2008;7(2):135–40.
89. de Rooij DG, Grootegoed JA. Spermatogonial stem cells. Curr Opin Cell Biol. 1998;10(6): 694–701.

90. Yoshida S. Stem cells in mammalian spermatogenesis. Dev Growth Differ. 2010;52(3):311–7.
91. Clermont Y. The cycle of the seminiferous epithelium in man. Am J Anat. 1963;112:35–51.
92. Clermont Y. Renewal of spermatogonia in man. Am J Anat. 1966;118(2):509–24.
93. Clermont Y. Spermatogenesis in man. A study of the spermatogonial population. Fertil Steril. 1966;17(6):705–21.
94. Clermont Y. Kinetics of spermatogenesis in mammals: seminiferous epithelium cycle and spermatogonial renewal. Physiol Rev. 1972;52(1):198–236.
95. Marshall GR, Plant TM. Puberty occurring either spontaneously or induced precociously in rhesus monkey (Macaca mulatta) is associated with a marked proliferation of Sertoli cells. Biol Reprod. 1996;54(6):1192–9.
96. He Z, Kokkinaki M, Dym M. Signaling molecules and pathways regulating the fate of spermatogonial stem cells. Microsc Res Tech. 2009;72(8):586–95.
97. Buaas FW, Kirsh AL, Sharma M, et al. Plzf is required in adult male germ cells for stem cell self-renewal. Nat Genet. 2004;36(6):647–52.
98. Costoya JA, Hobbs RM, Barna M, et al. Essential role of Plzf in maintenance of spermatogonial stem cells. Nat Genet. 2004;36(6):653–9.
99. Sada A, Suzuki A, Suzuki H, et al. The RNA binding protein NANOS2 is required to maintain murine spermatogonial stem cells. Science. 2009;325(5946):1394–8.
100. Suzuki H, Sada A, Yoshida S, et al. The heterogeneity of spermatogonia is revealed by their topology and expression of marker proteins including the germ cell-specific proteins Nanos2 and Nanos3. Dev Biol. 2009;336(2):222–31.
101. Spradling A, Drummond-Barbosa D, Kai T. Stem cells find their niche. Nature. 2001;414(6859):98–104.
102. Ohlstein B, Kai T, Decotto E, et al. The stem cell niche: theme and variations. Curr Opin Cell Biol. 2004;16(6):693–9.
103. Morrison SJ, Spradling AC. Stem cells and niches: mechanisms that promote stem cell maintenance throughout life. Cell. 2008;132(4):598–611.
104. Chiarini-Garcia H, Hornick JR, Griswold MD, et al. Distribution of type A spermatogonia in the mouse is not random. Biol Reprod. 2001;65(4):1179–85.
105. Chiarini-Garcia H, Raymer AM, Russell LD. Nonrandom distribution of spermatogonia in rats: evidence of niches in the seminiferous tubules. Reproduction. 2003;126(5):669–80.
106. Meng X, Lindahl M, Hyvonen ME, et al. Regulation of cell fate decision of undifferentiated spermatogonia by GDNF. Science. 2000;287(5457):1489–93.
107. Hofmann MC, Braydich-Stolle L, Dym M. Isolation of male germ-line stem cells; influence of GDNF. Dev Biol. 2005;279(1):114–24.
108. Naughton CK, Jain S, Strickland AM, et al. Glial cell-line derived neurotrophic factor-mediated RET signaling regulates spermatogonial stem cell fate. Biol Reprod. 2006;74(2): 314–21.
109. He Z, Jiang J, Hofmann MC, et al. Gfra1 silencing in mouse spermatogonial stem cells results in their differentiation via the inactivation of RET tyrosine kinase. Biol Reprod. 2007;77(4):723–33.
110. Tokuda M, Kadokawa Y, Kurahashi H, et al. CDH1 is a specific marker for undifferentiated spermatogonia in mouse testes. Biol Reprod. 2007;76(1):130–41.
111. Yoshinaga K, Nishikawa S, Ogawa M, et al. Role of c-kit in mouse spermatogenesis: identification of spermatogonia as a specific site of c-kit expression and function. Development. 1991;113(2):689–99.
112. Loveland KL, Schlatt S. Stem cell factor and c-kit in the mammalian testis: lessons originating from mother nature's gene knockouts. J Endocrinol. 1997;153(3):337–44.
113. Hermann BP, Sukhwani M, Hansel MC, et al. Spermatogonial stem cells in higher primates: are there differences from those in rodents? Reproduction. 2010;139(3):479–93.
114. Dym M, Kokkinaki M, He Z. Spermatogonial stem cells: mouse and human comparisons. Birth Defects Res C Embryo Today. 2009;87(1):27–34.
115. Hunt PA, Hassold TJ. Sex matters in meiosis. Science. 2002;296(5576):2181–3.
116. Morelli MA, Cohen PE. Not all germ cells are created equal: aspects of sexual dimorphism in mammalian meiosis. Reproduction. 2005;130(6):761–81.

117. Zhou Q, Li Y, Nie R, et al. Expression of stimulated by retinoic acid gene 8 (Stra8) and maturation of murine gonocytes and spermatogonia induced by retinoic acid in vitro. Biol Reprod. 2008;79:35–42.
118. Hermo L, Pelletier RM, Cyr DG, et al. Surfing the wave, cycle, life history, and genes/proteins expressed by testicular germ cells. Part 1: background to spermatogenesis, spermatogonia, and spermatocytes. Microsc Res Tech. 2010;73(4):241–78.
119. Cobb J, Handel MA. Dynamics of meiotic prophase I during spermatogenesis: from pairing to division. Semin Cell Dev Biol. 1998;9(4):445–50.
120. Lie PP, Cheng CY, Mruk DD. Coordinating cellular events during spermatogenesis: a biochemical model. Trends Biochem Sci. 2009;34(7):366–73.
121. Oakberg EF. Duration of spermatogenesis in the mouse and timing of stages of the cycle of the seminiferous epithelium. Am J Anat. 1956;99(3):507–16.
122. Heller CG, Clermont Y. Spermatogenesis in man: an estimate of its duration. Science. 1963; 140:184–6.
123. Don J, Stelzer G. The expanding family of CREB/CREM transcription factors that are involved with spermatogenesis. Mol Cell Endocrinol. 2002;187(1–2):115–24.
124. Foulkes NS, Mellstrom B, Benusiglio E, et al. Developmental switch of CREM function during spermatogenesis: from antagonist to activator. Nature. 1992;355(6355):80–4.
125. Foulkes NS, Schlotter F, Pevet P, et al. Pituitary hormone FSH directs the CREM functional switch during spermatogenesis. Nature. 1993;362(6417):264–7.
126. Delmas V, van der Hoorn F, Mellstrom B, et al. Induction of CREM activator proteins in spermatids: down-stream targets and implications for haploid germ cell differentiation. Mol Endocrinol. 1993;7(11):1502–14.
127. Nantel F, Monaco L, Foulkes NS, et al. Spermiogenesis deficiency and germ-cell apoptosis in CREM-mutant mice. Nature. 1996;380(6570):159–62.
128. Hogeveen KN, Sassone-Corsi P. Regulation of gene expression in post-meiotic male germ cells: CREM signalling pathways and male fertility. Hum Fertil (Camb). 2006;9(2):73–9.
129. Peri A, Serio M. The CREM system in human spermatogenesis. J Endocrinol Invest. 2000;23(9):578–83.
130. Miller D, Brinkworth M, Iles D. Paternal DNA packaging in spermatozoa: more than the sum of its parts? DNA, histones, protamines and epigenetics. Reproduction. 2010;139(2):287–301.
131. Gardiner-Garden M, Ballesteros M, Gordon M, et al. Histone- and protamine-DNA association: conservation of different patterns within the beta-globin domain in human sperm. Mol Cell Biol. 1998;18(6):3350–6.
132. Wykes SM, Krawetz SA. The structural organization of sperm chromatin. J Biol Chem. 2003;278(32):29471–7.
133. Moulson CL, Fong LG, Gardner JM, et al. Increased progerin expression associated with unusual LMNA mutations causes severe progeroid syndromes. Hum Mutat. 2007;28(9):882–9.
134. Arpanahi A, Brinkworth M, Iles D, et al. Endonuclease sensitive regions of human spermatozoal chromatin are highly enriched in promoter and CTCF binding sequences. Genome Res. 2009;19(8):1338–49.
135. Hammoud SS, Nix DA, Zhang H, et al. Distinctive chromatin in human sperm packages genes for embryo development. Nature. 2009;460(7254):473–8.
136. Dadoune JP. Spermatozoal RNAs: what about their functions? Microsc Res Tech. 2009;72(8): 536–51.
137. Mitchell RT, Saunders PT, Sharpe RM, et al. Male fertility and strategies for fertility preservation following childhood cancer treatment. Endocr Dev. 2009;15:101–34.
138. Morris ID. Sperm DNA damage and cancer treatment. Int J Androl. 2002;25(5):255–61.
139. Shalet SM, Tsatsoulis A, Whitehead E, et al. Vulnerability of the human Leydig cell to radiation damage is dependent upon age. J Endocrinol. 1989;120(1):161–5.
140. Stahl O, Eberhard J, Jepson K, et al. Sperm DNA integrity in testicular cancer patients. Hum Reprod. 2006;21(12):3199–205.

Chapter 4
Impact of Radiotherapy and Chemotherapy on the Testis

Carolina Ortega and Herman Tournaye

Thanks to improvements in therapeutic regimens, around 70–75% of children treated for cancer will survive and grow to adulthood [1]. It has been estimated that approximately 200,000 people living in the USA [2] and one in 715 people in the UK are childhood cancer survivors [3]. Approximately, half of childhood cancers are hematologic malignancies (leukemia and lymphoma) and the anticipated long-term survival for children with these disorders is nowadays exceeding 75% [4]. Testicular cancer represents about 1% of all cancers in men, but it is the most common solid organ tumor between 25 and 35 years of age with an incidence of 3–6 new cases occurring per 100,000 males per year in Western society and up to 9–10 new cases per year in Germany and Scandinavian countries [5]. The use with chemotherapy and radiotherapy combined with surgical techniques has enabled the cure of 90% of the cases [6] to the extent that testicular cancer therapy is one of the outstanding successes in medicine today [7–9].

Unfortunately, during cancer treatment different organ systems are affected, including cardiac, pulmonary, renal, hepatic, and endocrine system. Since the quality of life has become an important factor, one of the major concerns of patients treated for cancer is infertility related to either the malignancy itself or more frequently, the treatment.

Adverse Effects on the Testis as an Endocrine Organ

The testis is highly susceptible to the toxic effects of radiation and chemotherapy at all ages of life. It is one of the most radiosensitive tissues, and damage can be caused by direct irradiation or from scattered irradiation at adjacent tissues [10].

H. Tournaye, M.D., Ph.D. (✉)
Centre for Reproductive Medicine, University Hospital, Dutch-speaking Brussels Free University, Laarbeeklaan 101, Brussels 1090, Belgium
e-mail: tournaye@uzbrussel.be

E. Seli and A. Agarwal (eds.), *Fertility Preservation in Males: Emerging Technologies and Clinical Applications*, DOI 10.1007/978-1-4614-5620-9_4,
© Springer Science+Business Media New York 2012

After chemotherapy, testicular damage is drug specific and dose related [11, 12]. The age, at which chemotherapy and radiotherapy are carried out, is equally important. It has been suggested that the germinal epithelium of the prepubertal testis is less susceptible to damage than the adult testis [13]. However, if the doses of chemotherapy are calculated on a per square meter and calculating radiation doses to the gonad given during childhood and adolescence, some chemotherapy agents and radiotherapy doses that produce nonreversible azoospermia in those patients, seems to be the same as those for adults [14]. Two important endocrinologically active cells are found in the testis, i.e., the Sertoli cells (SCs) and the Leydig cells (LCs).

Effects on Sertoli Cell Function

The SC has a specific role during fetal life in the testicular development and sexual differentiation by secreting anti-Mullerian hormone which ensures the regression of the Mullerian ducts and by secreting testosterone to develop a male phenotype. The SC proliferates during two specific periods of life, i.e., during the fetal and neonatal period (up to 12–18 months postpartum) and during the pubertal period. After the neonatal period, the SC enters into a quiescent period which changes into an active replicating period at puberty in order to support further spermatogenesis during adult life [15]. During puberty, SC switches from an immature proliferative stage to a mature nonproliferative stage. Adult testis size and daily spermatozoa production are related to the number of SCs determining as such the number of germ cells that can be produced [16, 17].

Cytotoxic treatments can only cause damage to the SC population when these cells are replicating, and as a result in adults SCs are radio- and chemotherapy resistant [16, 18].

Yet in a recent case report, possible Sertoli cell damage because of chemotherapy has been reported [19]. The authors demonstrated that a patient who underwent chemotherapy during puberty shows SC immunopositive for CK-18 during adulthood in approximately 13% of SC in tubules with impaired spermatogenesis. While SCs are relatively resistant to chemotherapy, the presence of CK-18 may reflect a dedifferentiate of SC as a consequence of testicular damage produced by chemotherapy treatment.

Effects on Leydig Cell Function

The LC produces both testosterone and insulinlike factor 3, playing a role in downstream events of masculinization and in the descent of the testes into the scrotum in fetal life [20, 21]. At puberty, the production of testosterone increases thereby inducing the secondary sex differentiation. Testosterone output then gradually decreases from age 30 [22].

Chemotherapy may have a direct toxic effect on the LC. It has been reported that after chemotherapy 59–75% of men have increased luteinizing hormone (LH) levels in the presence of low or normal testosterone [23–26]. Some of these men with subnormal testosterone levels may also show reduced bone density, body composition, and disturbance of quality of life [27, 28]. However, there is an evidence of recovery of LC function during the first 10 years after chemotherapy [29].

Regarding the effect of radiotherapy, LC function may be more prone to damage from irradiation in prepubertal life compared with adulthood [30], Patients undergoing single dose of above 0.75 Gy [13] produce significant rises in LH which is also seen with fractionated doses of radiation of above 2 Gy [31], although values gradually return to normal levels over 30 months and testosterone values are not altered with these doses [28]. However, doses above 30 Gy produce significant decrease in testosterone serum values [30].

Adverse Effects on the Testis as an Exocrine Organ

Spermatogenesis, the exocrine function of the testis, starts at puberty and continues throughout adult life. Pluripotent spermatogonial stem cell undergoes self-renewal, apoptosis, and differentiation into mature spermatozoa [17]. On average, the total number of germ cells during childhood varies from 13 to 83×10^6 [32]. Two different populations of spermatogonial stem cells exist, i.e., the A_{dark} which are reserve stem cells with a low mitotic activity, and the A_{pale} which are the true male germline progenitor cells with a high mitotic activity. A_{pale} can differentiate into B spermatogonia around 4–5 years of age after which their number slowly increases until the age of 8–9 years. While type B spermatogonia are being produced, the true stem cell population gets refueled since half of the sister cells remain A_{pale} spermatogonia [33]. Then, a marked spermatogonial proliferation period starts. The development of the testes can lead to postpubertal sperm counts of up to several 100 million (or more) sperm per ml of semen depending on the adult testicular volume reached [34].

Effects of Radiotherapy on the Exocrine Function

Radiotherapy can be applied either by electromagnetic radiation (X-rays) and corpuscular radiation (electrons) produced by a linear accelerator or by rays generated by the decay of the cobalt 60 radioisotope. It is well documented, that impairment of spermatogenesis following single dose radiation depends on the dose and the age in which the patient underwent radiation therapy but impairment on the number of spermatozoa amount is always observed: oligozoospermia can be observed with doses as low as 0.1 Gy due to damage at the spermatogonial stage while azoospermia is observed when the patient is irradiated with doses above 2 Gy. With doses of 2–3 Gy, the spermatocytes are being damaged and azoospermia is observed 60 days after the onset of the treatment and at doses of 4–6 Gy, damage occurs up to the spermatid stage, and thus azoospermia is observed earlier [13]. Recovery of spermatogenesis is observed depending on the total dose of irradiation, on the survival of the spermatogonial stem cells of both "true" and "reserve" compartments and their ability to further differentiate. Hence, a complete recovery of sperm concentration is seen after 9–18 months with doses below 1 Gy [35], after 30 months with doses between 2 and 3 Gy and after 5 years with doses above 4 Gy [13]. However, doses above 20 Gy causing also LC damage are irreversible [35].

Many patients, however, will undergo fractionated radiotherapy which induces more gonadal toxicity compared to single-dose radiotherapy. Several studies showed that patients who received fractionated irradiation with doses between 1 and 3 Gy became azoospermic, and the recovery of spermatogenesis following the treatment was observed only in few patients [36–38]. Radiation doses of 1.2 Gy are considered as a threshold for permanent testicular damage without recovery in most of the patients. Yet in patients with spermatogenesis recovering after radiotherapy, the frequencies of numerical and structural chromosomal abnormalities were found to be increased in relation to the radiation dose used [39].

Total body irradiation is associated with even higher gonadal toxicity. Eighty percent of patients who underwent TBI after stem cell transplantation will have permanent gonadal failure [40] with only 0.5–1% of patients who received, respectively, 12 and 10 Gy or more during TBI, achieving fatherhood [41].

Effects of Chemotherapy on the Exocrine Function

Nowadays, different types of chemotherapeutical agents are used for cancer treatment. They interfere with DNA and RNA synthesis, inhibit protein synthesis or prevent microtubule functions, all vital for successful cell division. Effects of chemotherapy depend on the dose and the agent used with alkylating agents being most aggressive, e.g., procarbazine and cyclophosphamide.

Testicular damage caused by cytotoxic drugs was first described in 1948, when azoospermia was reported in 27 of 30 men following the treatment with nitrogen mustard [42]. The use of chemotherapy and radiotherapy produce a combination of destruction of the proliferating germ cell pool and inhibition of further differentiation of the survival germs cells. Different radiotherapy and chemotherapy regimens vary widely in their effect on fertility and can lead to effects from temporary oligozoospermia to permanent azoospermia. Damage to the other aspects of sperm function, such as loss of motility or morphological abnormalities, are less pronounced. If spermatozoa are produced after therapy, their motility and percentage that exhibit normal morphology are restored to pretreatments levels [43, 44]. However, 15–30% remains sterile in the long term [45]. The American Society for Clinical Oncology has published a document including an overview of the anticipated sperm quality after various chemotherapeutic drugs and guidelines on fertility preservation promoting cryobanking whenever possible [46]. A working party of the UK Royal Colleges in 2007 recommended that all patients who required anticancer treatment should be fully informed about the potential gonadotoxic side effects at the time of the diagnosis and before potentially gonadotoxic treatments [47]. Unfortunately, up to 40% of adult patients are not informed about the sterilizing side effects of their treatment [48] and the same figure is reported in children cancer [49]. The latter survey also shows that in the subpopulation of children at high risk for sterility only in 14% of cases the problem of sterility is discussed with the parents [49].

When cytotoxic drugs affect replicating spermatogonial stem cells, DNA damage can probably be repaired and the risk for transmissible mutations may be low. The cytotoxic agents also affect the differentiating germinal cells. If the damage is produced at the spermatocyte stage, DNA damage can be repaired. When chemotherapy affects at spermatids or testicular spermatozoa transmissible gene mutations and chromosomal aberrations can result because at these stages, DNA damage cannot be repaired [14].

Pharmacological Prevention

Methods to prevent the sterilizing side effects of radio- and chemotherapy on sperm production and improve gonadal function after cancer treatment are extremely important and now more emphasis is put on the quality of life after the cancer cure.

Several biochemical and biological approaches have been studied in animal models to protect the testis against cytotoxic agents [50]. These methods focus on hormonal modulation to prevent or reverse damage to the germ line during cancer treatment. This strategy is based on the observation that in rodents, germinal epithelium cells are more resistant to cytotoxic agents in their nonreplicating state rather than in their proliferative state. Suppression therapy with gonadotropin and testosterone in rats before or after cytotoxic agents considerably improve spermatogenesis and fertility [51]. Some clinical trials have been performed in men using suppression hormonal therapy. However, this treatment in men did not improve sperm count and fertility as in rats [52–55]. Hence, the application of these procedures to humans remains uncertain and more studies must be done.

Prevention by Gamete Preservation

A large number of cancer patients survive after radio and chemotherapy treatments. These patients expect the quality of life, including fatherhood. The American Society of Clinical Oncology (ASCO) recommends that fertility treatments must be discussed with patients and advices depend on the patient's age, disease, prognosis, and time interval before treatment. Hence, all patients who undergo anticancer treatment may be considered for cryobanking. Semen cryopreservation should be performed preferentially before starting any treatment. However, semen sample can be taken even when treatment has already started [56]. One single semen sample, even of limited quality, is sufficient to perform several intracytoplasmic sperm injections (ICSIs) cycles and urgency for starting chemotherapy has become an unlikely excuse.

Semen cryostorage has been shown to be an effective strategy and is recommended by the National Institute for Health and Clinical Excellence (NICE) and ASCO [46, 57]. In adolescent boys, sperm can be obtained by masturbation [58].

In the situation of ejaculation failure, the search for spermatozoa in a urine sample could be proposed [59, 60] and eventually semen can be obtained using penile vibratory stimulation or rectal electrostimulation technique under general anesthesia. When spermatozoa cannot be obtained using the techniques described above, a fine needle aspiration (FNA) or testicular sperm extraction (TESE) can be performed.

In prepubertal boys having immature testis, there are no options currently available for fertility preservation at diagnosis and options remain experimental. Testicular biopsy (TESE) and spermatogonial stem cell cryopreservation might be an option for those prepubertal patients [61, 62].

In a mouse model, it has been shown that testicular germ cells transplanted into a sterile mice, can restore fertility and reproduction in vivo [63, 64].

One of the mayor concerns, however, is the possibility of contamination of testicular tissue or spermatogonial stem cells that could be transplanted back and reexposing him to cancer later on [65]. Nowadays, there is no safe technology available preventing transmission of malignant cells by this novel strategy [66].

Attitudes of Physicians, Parents, and Patients Toward Preservation

Fertility problems have a major impact on the future quality of life of patients and physicians and parents are aware of it. However, discussing the storage of sperm of an adolescent with cancer is a challenging aspect of pediatric oncology care. A study carried out in The Netherlands showed that physicians always wanted a separated discussion with the adolescent because of the sensitive nature of the topic [67]. Ginsberg et al. demonstrated that parents of prepubertal boys are willing to pursue testicular cryopreservation at diagnosis even if the diagnosis is stressful. Factors, such as religion, finance, ethics and experimental nature of cryopreservation, did not play a major role in decision-making [62]. Concerning child participation during the decision-making process, a study carried out by de Vries et al. parents and physicians do not have the same opinion. All physicians agree that children should participate in the decision-making; however, some parents doubted whether the issue should be discussed with their son [67]. And although children can have some ideas about the effects of cancer treatment on them fertility, normally their parents are more worried about preserving reproductive function. Regarding the in vitro techniques that can be used in patients who have been treated for cancer, a questionnaire sent to oncologists in Minnesota showed that 74% of the responding oncologists were unaware of recent advances in assisted reproduction and did not know about the existence of ICSI [68] and an Irish survey corroborates these findings [69]. Unfortunately, oncologists offer semen cryopreservation to <25% of their adolescent patients [48]. Risk of fertility is only discussed with 76% of post-pubertal and 61% of prepubertal boys and in most cases discussion occurred at diagnosis, but in 7% it is delayed until during treatment [49].

Ethical, Legal, and Economic Considerations for Gamete Preservation

Patients must be informed about fertility preservation, and the need for an informed consent is mandatory. Fertility preservation in children causes a difficult ethical problem. The informed consent concerning children needs the involvement of their parents or legal guardians. When the patient is able to understand the issue (postpu-bertal boys), he can give an assent (permission less than full consent) together with parents consent. Regarding the TESE in adolescents, an ethical board approval may be needed before considering surgery. Patients must be informed about the financial cost of cryopreservation and know that in some cases insurances will not cover all the costs [70].

Pregnancy and Assisted Reproduction Techniques After Cancer Treatment

The introduction of ICSI has totally changed the reproductive prospects for boys and men who underwent radiotherapy and chemotherapy. Hence, cured cancer patients can now father children who are genetically their own, even with the poor-est semen samples [71–77]. A study carried out by Naysmith, showed that 45% of the patients after anticancer treatment could potentially benefit from ICSI [74]. In sexually mature men, frozen-thawed spermatozoa obtained by TESE before cyto-toxic treatments can be successfully used performing ICSI [78, 79] even if the sperm has extremely poor characteristics of count, motility, and morphology [61]. In pre-pubertal boys, cryopreservation of mature and immature sperm before starting anti-cancer treatment is feasible. Once the child is cured from the cancer, thawed testicular tissue can be transplanted back into his own testis or can be matured in vitro or in vivo until the stage at which he will be competent to procure normal fertilization with ICSI.

Regarding the possible congenital malformations in children of patients who underwent anticancer treatments, some authors showed that the risk is negligible [1, 80–83] though other authors demonstrated a measurable increase in the frequency of sperm chromosomal abnormalities related to cytotoxic treatments [84, 85]. The risk of producing chromosomally abnormal sperm is highest in the few weeks after completion of chemotherapy, so men should wait for 6 months after stopping their chemotherapy before conception is attempted [86]. It has been demonstrated that men treated with chemotherapy for Hodgkin's disease or testicular cancer had an increased aneuploidy in their spermatozoa [87–89], Patients with unilateral testicular cancer show a high prevalence of chromosomal abnormalities of spermatogenesis, which might be observed in contralateral testis [90] because of the presence of the cancer itself. However, children fathered by patients treated for cancer, did not show more congenital disorders [82]. Hence, given the absence of any clinical evidence of chromosomal abnormalities in offspring born from men undergoing chemotherapy or men who had chemotherapy, even patients who already started their cancer treatment can be offered cryobanking [56]. Additionally, in order to control for aneuploidy, preimplantation genetic diagnosis (PGD) may be suggested whenever assisted reproduction is performed using sperm banked during chemotherapy.

Conclusions

Currently, the effectiveness of anticancer treatments has improved the survival rates tremendously and as a result most of the patients are cured. The quality of life has now become an important issue, including the posttreatment fertility status. However, depending on the dose and type of drugs used, radiotherapy and chemotherapy can produce gonadal toxicity ranging from temporary oligozoospermia to permanent azoospermia.

The testis has both an endocrine and an exocrine function. The endocrine output depends on an adequate LC function. For the exocrine function, i.e., spermatogenesis, both a functional SC and a spermatogonial stem cell pool are necessary. Spermatogonial stem cells are undifferentiated cells that give rise to the spermatogenic cells and, finally, the spermatozoa. Even though there is a continual loss of differentiated cells, the spermatogenic cell lineage maintains its cell number; thanks to the adult spermatogonial stem cells, which produce both new stem cells (self-renewal) and differentiating cells.

LCs are rather resistant to chemo- and radiotherapy. While elevated LH indicates LC dysfunction, most men will show normal testosterone levels and will retain a normal bone density. While routine testosterone supplementation after gonadotoxic treatment is not indicated, men should have endocrine monitoring after cancer treatment and in selected cases testosterone substitution may be required. LCs seem more vulnerable to chemo and radiotherapy in prepubertal life.

Also SCs show a good resistance to chemo and radiotherapy, except in the post-natal and pubertal period when their numbers are increasing through mitosis. Increases in FSH are due to indirect effects through loss of the germ cell pool.

The extent of the gonadotoxicity on this germ cell is strongly related to the nature of the specific agents that are used and their dose or to the intensity of the radiation and the place of the body where it was administered. Transient reductions of sperm count can occur even after mild forms of chemotherapy or low doses of gonadal radiation, due to the destruction of the sensitive differentiating spermatogonia. Stronger chemotherapeutic regimes or higher doses of gonadal irradiation, however, lead to prolonged reduction in sperm count or complete azoospermia. Whether sperm production will eventually recover depends on the survival of the spermatogonial stem cells and the integrity of their ability to differentiate. High doses of radiation to the testis (>2.5 Gy) cause DNA damage and cell death. The most damaging chemotherapeutic agents in the adult man are the alkylating agents.

Even though the gonadotoxic effect of most anticancer treatments is known, it is difficult to predict the final effect on the fertility potential of the patient. There remain important interindividual differences in response to the treatment and even if a regimen with low gonadotoxicity is started, it is possible that eventually a more gonadotoxic treatment has to be administered because of earlier treatment failure.

Patients undergoing anticancer therapy must be informed about their future fertility status and oncologist must discuss at diagnosis the different options available to preserve fertility. If the patient is a child, this information should be given to the parents or legal guardians.

Cryobanking should be offered and performed preferentially before starting any treatment. Otherwise, semen cryopreservation during the first months of treatment can be considered. When the patient is prepubertal, spermatogonial stem cell banking, although still being in a preclinical phase of development, can be considered.

References

1. Grundy R, Gosden RG, Hewitt M, et al. Fertility preservation for children treated for cancer (1): scientific advances and research dilemmas. Arch Dis Child. 2001;84(4):355–9. Review.
2. Hollen PJ, Hobbie WL. Establishing comprehensive specialty follow-up clinics for long-term survivors of cancer. Providing systematic physiological and psychosocial support. Support Care Cancer. 1995;3(1):40–4.
3. SIGN 76. The long-term follow up of children treated for cancer. www.sign.com. Accessed Nov 2004.
4. Fallat ME, Hutter J, American Academy of Pediatrics Committee on Bioethics, et al. Preservation of fertility in pediatric and adolescent patients with cancer. Pediatrics. 2008;121(5):e1461–9. Review.
5. Huyghe E, Matsuda T, Thonneau P. Increasing incidence of testicular cancer worldwide: a review. J Urol. 2003;170(1):5–11. Review.
6. Fosså SD, Abyholm T, Normann N, et al. Posttreatment fertility in patients with testicular cancer. III. Influence of radiotherapy in seminoma patients. Br J Urol. 1986;58(3):315–9.
7. Presti JC, Herr HW, Carroll PR. Fertility and testis cancer. Urol Clin North Am. 1993;20(1):173–9. Review.

8. Dearnaley D, Huddart R, Horwich A. Regular review: managing testicular cancer. BMJ. 2001;322(7302):1583–8. Review.
9. Laguna MP, Pizzocaro G, Klepp O, et al. EAU guidelines on testicular cancer. Eur Urol. 2001;40(2):102–10.
10. Howell SJ, Shalet SM. Spermatogenesis after cancer treatment: damage and recovery. J Natl Cancer Inst Monogr. 2005;34:12–7. Review.
11. Pryzant RM, Meistrich ML, Wilson G, et al. Longterm reduction in sperm count after chemotherapy with and without radiation therapy for non-Hodgkin's lymphomas. J Clin Oncol. 1993;11(2):239–47.
12. da Cunha MF, Meistrich ML, Fuller LM, et al. Recovery of spermatogenesis after treatment for Hodgkin's disease: limiting dose of MOPP chemotherapy. J Clin Oncol. 1984;2(6):571–7.
13. Rowley MJ, Leach DR, Warner GA, et al. Effect of graded doses of ionizing radiation on the human testis. Radiat Res. 1974;59(3):665–78.
14. Meistrich ML. Male gonadal toxicity. Pediatr Blood Cancer. 2009;53(2):261–6. Review.
15. Sharpe RM, McKinnell C, Kivlin C, et al. Proliferation and functional maturation of Sertoli cells, and their relevance to disorders of testis function in adulthood. Reproduction. 2003;125(6):769–84. Review.
16. Orth JM, Gunsalus GL, Lamperti AA. Evidence from Sertoli cell-depleted rats indicates that spermatid number in adults depends on numbers of Sertoli cells produced during perinatal development. Endocrinology. 1988;122(3):787–94.
17. Sharpe RM. Regulation of spermatogenesis. In: Knobil E, Neill JD, editors. The physiology of reproduction. 2nd ed. New York: Raven; 1994. p. 1363–434.
18. Jégou B, Sharpe RM. Paracrine mechanisms in testicular control. In: de Kretser DM, editor. Molecular biology of the male reproductive system. New York: Academic; 1993. p. 271–310.
19. Bar-Shira Maymon B, Yogev L, Marks A, et al. Sertoli cell inactivation by cytotoxic damage to the human testis after cancer chemotherapy. Fertil Steril. 2004;81(5):1391–4.
20. Hutson JM, Hasthorpe S, Heyns CF. Anatomical and functional aspects of testicular descent and cryptorchidism. Endocr Rev. 1997;18(2):259–80. Review.
21. Sharpe RM. Hormones and testis development and the possible adverse effects of environmental chemicals. Toxicol Lett. 2001;120(1–3):221–32. Review.
22. Vermeulen A. The male climaterium. Ann Med. 1993;25(6):531–4.
23. Howell SJ, Radford JA, Ryder WD, et al. Testicular function after cytotoxic chemotherapy: evidence of Leydig cell insufficiency. J Clin Oncol. 1999;17(5):1493–8.
24. Hansen SW, Berthelsen JG, von der Maase H. Longterm fertility and Leydig cell function in patients treated for germ cell cancer with cisplatin, vinblastine, and bleomycin versus surveillance. J Clin Oncol. 1990;8(10):1695–8.
25. Palmieri G, Lotrecchiano G, Ricci G, et al. Gonadal function after multimodality treatment in men with testicular germ cell cancer. Eur Endocrinol. 1996;134(4):431–6.
26. Stuart NS, Woodroffe CM, Grundy R, et al. Longterm toxicity of chemotherapy for testicular cancer – the cost of cure. Br J Cancer. 1990;61(3):479–84.
27. Holmes SJ, Whitehouse RW, Clark ST, et al. Reduced bone mineral density in men following chemotherapy for Hodgkin's disease. Br J Cancer. 1994;70(2):371–5.
28. Howell SJ, Shalet SM. Effect of cancer therapy on pituitary-testicular axis. Int J Androl. 2002;25(5):269–76. Review.
29. Howell SJ, Radford JA, Ryder WD, Shalet SM. Testicular function after cytotoxic chemotherapy: evidence of Leydig cell insufficiency. J Clin Oncol. 1999;17(5):1493–8.
30. Shalet SM, Tsatsoulis A, Whitehead E, et al. Vulnerability of the human Leydig cell to radiation damage is dependent upon age. J Endocrinol. 1989;120(1):161–5.
31. Shapiro E, Kinsella TJ, Makuch RW, et al. Effects of fractionated irradiation of endocrine aspects of testicular function. J Clin Oncol. 1985;3(9):1232–9.
32. Muller J, Skakkebaek NE. Quantification of germ cells and seminiferous tubules by stereological examination of testicles from 50 boys who suffered from sudden death. Int J Androl. 1983;6(2):143–56.

33. Ehmcke J, Wistuba J, Schlatt S. Spermatogonial stem cells: questions, models and perspectives. Hum Reprod Update. 2006;12(3):275–82.
34. World Health Organisation. WHO manual for the standardized investigation, diagnosis and management of the infertile male. Cambridge: Cambridge University Press; 2000.
35. Colpi GM, Contalbi GF, Nerva F, et al. Testicular function following chemo-radiotherapy. Eur J Obstet Gynecol Reprod Biol. 2004;113 Suppl 1:S2–6. Review.
36. Speiser B, Rubin P, Casarett G. Aspermia following lower truncal irradiation in Hodgkin's disease. Cancer. 1973;32(3):692–8.
37. Centola GM, Keller JW, Henzler M, et al. Effect of low-dose testicular irradiation on sperm count and fertility in patients with testicular seminoma. J Androl. 1994;15(6):608–13.
38. Hahn EW, Feingold SM, Nisce L. Aspermia and recovery of spermatogenesis in cancer patients following incidental gonadal irradiation during treatment: a progress report. Radiology. 1976;119(1):223–5.
39. Martin RH, Hildebrand K, Yamamoto J, et al. An increased frequency of human sperm chromosomal abnormalities after radiotherapy. Mutat Res. 1986;174:219–25.
40. Socie G, Salooja N, Cohen A, et al. Late effects working party of the European study group for blood and marrow transplantation. Nonmalignant late effects after allogeneic stem cell transplantation. Blood. 2003;101:3373–85.
41. Sanders JE, Hawley J, Levy W, et al. Pregnancies following high-dose cyclophosphamide with or without high-dose busulfan or total body irradiation and bone marrow transplantation. Blood. 1996;87:3045–52.
42. Spitz S. The histological effects of nitrogen mustards on human tumors and tissues. Cancer. 1948;1(3):383–98.
43. Meistrich ML, Wilson G, Brown BW, et al. Impact of cyclophosphamide on long-term reduction in sperm count in men treated with combination chemotherapy for Ewing and soft tissue sarcomas. Cancer. 1992;70(11):2703–12.
44. Gandini L, Sgro P, Lombardo F, et al. Effect of chemo- or radiotherapy on sperm parameters of testicular cancer patients. Hum Reprod. 2006;21(11):2882–9.
45. Schrader M, Muller M, Straub B, et al. The impact of chemotherapy on male fertility: a survey of the biologic basis and clinical aspects. Reprod Toxicol. 2001;15(6):611–7. Review.
46. Lee SJ, Schover LR, Partridge AH, et al. American Society of Clinical Oncology recommendations on fertility preservation in cancer patients. J Clin Oncol. 2006;24(18):2917–31. Erratum in: J Clin Oncol. 2006;24(36):5790.
47. Royal College of Physicians, Royal College of Radiologist, Royal College of Obstetricians and Gynaecologist. The effects of cancer treatment on reproductive functions. Guidance on management. Report of a working party. 2007. http://www.rcog.org.uk/resources/public/pdf/EffectCancerReprod.pdf.
48. Schover LR, Brey K, Lichtin A, et al. Oncologists' attitudes and practices regarding banking sperm before cancer treatment. J Clin Oncol. 2002;20(7):1890–7.
49. Anderson RA, Weddell A, Spoudeas HA, et al. Do doctors discuss fertility issues before they treat young patients with cancer? Hum Reprod. 2008;23(10):2246–51.
50. Meistrich ML, Zhang Z, Porter KL, Bolden-Tiller OU, Shetty G. Prevention of adverse effects of cancer treatment on the germline. In: Anderson D, Brinkworth MH, editors. Male-mediated developmental toxicity. Cambridge: Royal Society of Chemistry; 2007. p. 114–23.
51. Shetty G, Meistrich M. Hormonal approaches to preservation and restoration of male fertility after cancer treatment. J Natl Cancer Inst Monogr. 2005;34:36–9. Review.
52. Johnson DH, Linde R, Hainsworth JD, et al. Effect of a luteinizing hormone releasing hormone agonist given during combination chemotherapy on post-therapy fertility in male patients with lymphoma: preliminary observations. Blood. 1985;65:832–6.
53. Waxman JH, Ahmed R, Smith D, et al. Failure to preserve fertility in patients with Hodgkin's disease. Cancer Chemother Pharmacol. 1987;19:159–62.
54. Redman JR, Bajorunas DR. Suppression of germ cell proliferation to prevent gonadal toxicity associated with cancer treatment. In: Workshop on psychosexual and reproductive issues affecting patients with cancer. New York: American Cancer Society; 1987. p. 90–4.

55. Masala A, Faedda R, Alagna S, et al. Use of testosterone to prevent cyclophosphamide-induced azoospermia. Ann Intern Med. 1997;126:292–5.
56. Carson SA, Gentry WL, Smith AL, et al. Feasibility of semen collection and cryopreservation during chemotherapy. Hum Reprod. 1991;6(7):992–4.
57. National Institute for Health and Clinical Excellence. Fertility assessment and treatment for people with fertility problems. 2004. www.rcog.org.uk/resources/Public/pdf/Fertility_summary.pdf.
58. Kliesch S, Behre HM, Jurgens H, et al. Cryopreservation of semen from adolescent patients with malignancies. Med Pediatr Oncol. 1996;26(1):20–7.
59. Bahadur G, Ling KL, Hart R, et al. Semen production in adolescent cancer patients. Hum Reprod. 2002;17(10):2654–6.
60. Menon S, Rives N, Mousset-Simeon N, et al. Fertility preservation in adolescent males: experience over 22 years at Rouen University Hospital. Hum Reprod. 2009;24(1):37–44.
61. Tournaye H, Goossens E, Verheyen G, et al. Preserving the reproductive potential of men and boys with cancer: current concepts and future prospects. Hum Reprod Update. 2004;10(6):525–32. Review.
62. Ginsberg JP, Carlson CA, Lin K, Hobbie WL, et al. An experimental protocol for fertility preservation in prepubertal boys recently diagnosed with cancer: a report of acceptability and safety. Hum Reprod. 2010;25(1):37–41.
63. Brinster RL, Zimmermann JW. Spermatogenesis following male germ-cell transplantation. Proc Natl Acad Sci U S A. 1994;91(24):11298–302.
64. Brinster RL, Avarbock MR. Germline transmission of donor haplotype following spermatogonial transplantation. Proc Natl Acad Sci U S A. 1994;91(24):11287–9.
65. Jahnukainen K, Hou M, Petersen C, et al. Intratesticular transplantation of testicular cells from leukemic rats causes transmission of leukemia. Cancer Res. 2001;61(2):706–10.
66. Geens M, Van de Velde H, De Block G, et al. The efficiency of magnetic-activated cell sorting and fluorescence-activated cell sorting in the decontamination of testicular cell suspensions in cancer patients. Hum Reprod. 2007;22(3):733–42.
67. de Vries MC, Bresters D, Engberts DP, et al. Attitudes of physicians and parents towards discussing infertility risks and semen cryopreservation with male adolescents diagnosed with cancer. Pediatr Blood Cancer. 2009;53(3):386–91.
68. Zapzalka DM, Redmon JB, Pryor JL. A survey of oncologists regarding sperm cryopreservation and assisted reproductive techniques for male cancer patients. Cancer. 1999;86:1812–7.
69. Allen C, Keane D, Harrison RF. A survey of Irish consultants regarding awareness of sperm freezing and assisted reproduction. Ir Med J. 2003;96(1):23–5.
70. Deepinder F, Agarwal A. Technical and ethical challenges of fertility preservation in young cancer patients. Reprod Biomed Online. 2008;16(6):784–91.
71. Chen SU, Ho HN, Chen HF, et al. Pregnancy achieved by intracytoplasmic sperm injection using cryopre- served semen from a man with testicular cancer. Hum Reprod. 1996;11(12):2645–7.
72. Hallak J, Sharma RK, Thomas Jr AJ, et al. Why cancer patients request disposal of cryopreserved semen specimens posttherapy: a retrospective study. Fertil Steril. 1998;69(5):889–93.
73. Lass A, Akagbosu F, Abusheikha N, et al. A programme of semen cryopreservation for patients with malignant disease in a tertiary infertility centre: lessons from 8 years' experience. Hum Reprod. 1998;13(11):3256–61.
74. Naysmith TE, Blake DE, Harvey VJ, et al. Do men undergoing sterilizing cancer treatments have a fertile future? Hum Reprod. 1998;13(11):3250–5.
75. Tournaye H. Storing reproduction for oncological patients. Mol Cell Endocrinol. 2000;169(1–2):133–6. Review.
76. Ginsburg ES, Yanushpolsky EH, Jackson KV. In vitro fertilization for cancer patients and survivors. Fertil Steril. 2001;75(4):705–10.
77. Horne G, Atkinson A, Brison DR, et al. Achieving pregnancy against the odds: successful implantation of frozen-thawed embryos generated by ICSI using spermatozoa banked prior to chemo/radiotherapy for Hodgkin's disease and acute leukaemia. Hum Reprod. 2001;16(1): 107–9.

78. Hovatta O, Foudila T, Siegberg R, et al. Pregnancy resulting from intracytoplasmic injection of spermatozoa from a frozen-thawed testicular biopsy specimen. Hum Reprod. 1996;11(11): 2472–3.
79. Oates RD, Mulhall J, Burgess C, et al. Fertilization and pregnancy using intentionally cryopreserved testicular tissue as the sperm source for intracytoplasmic sperm injection in 10 men with non-obstructive azoospermia. Hum Reprod. 1997;12(4):734–9.
80. Byrne J, Mulvihill JJ, Connelly RR, et al. Reproductive problems and birth defects in survivors of Wilms' tumor and their relatives. Med Pediatr Oncol. 1988;16(4):233–40.
81. Senturia YD, Peckham CS. Children fathered by men treated with chemotherapy for testicular cancer. Eur J Cancer. 1990;26(4):429–32.
82. Hawkins MM, Draper GJ, Smith RA. Cancer among 1,348 offspring of survivors of childhood cancer. Int J Cancer. 1989;43(6):975–8.
83. Li FP, Fine W, Jaffe N, et al. Offspring of patients treated for cancer in childhood. J Natl Cancer Inst. 1979;62(5):1193–7.
84. Brandriff BF, Meistrich ML, Gordon LA, et al. Chromosomal damage in sperm of patients surviving Hodgkin's disease following MOPP (nitrogen mustard, vincristine, procarbazine, and prednisone) therapy with and without radiotherapy. Hum Genet. 1994;93(3):295–9.
85. Jenderny J, Rohrborn G. Chromosome analysis of human sperm. I. First results with a modified method. Hum Genet. 1987;76(4):385–8.
86. Revel A, Revel-Vilk S. Pediatric fertility preservation: is it time to offer testicular tissue cryopreservation? Mol Cell Endocrinol. 2008;282(1–2):143–9. Review.
87. Genesca A, Benet J, Caballin MR, et al. Significance of structural chromosome aberrations in human sperm: analysis of induced aberrations. Hum Genet. 1990;85(5):495–9.
88. Monteil M, Rousseaux S, Chevret E, et al. Increased aneuploid frequency in spermatozoa from a Hodgkin's disease patient after chemotherapy and radiotherapy. Cytogenet Cell Genet. 1997;76(3–4):134–8.
89. Robbins WA, Meistrich ML, Moore D, et al. Chemotherapy induces transient sex chromosomal and autosomal aneuploidy in human sperm. Nat Genet. 1997;16(1):74–8.
90. Gurgan T, Salman C, Demirol A. Pregnancy and assisted reproduction techniques in men and women after cancer treatment. Placenta. 2008;29(Suppl B):152–9. Review.

Chapter 5
Impact of Paternal Exposure to Gonadotoxins on Embryo and Offspring and the Male Evaluation

Kathleen Hwang, Paul Gittens, Desiderio Avila Jr., and Larry I. Lipshultz

The failure to conceive within 1 year occurs in approximately 15% of couples [1]. Approximately 50% of problems related to conception is either caused entirely by the male or is a combined problem with the male and his partner.

Gonadotoxic Exposure

There are several routes of exposure to gonadotoxins for the male. Exposures originate from lifestyle choices, medications, and treatment for malignancies, as well as occupational and chemical hazards. Regardless of the route, the impact of paternal exposure to gonadotoxins on subsequent fertility is significant as it not only affects the individual, but may also affect future offspring.

Gonadotoxic Effects of Cancer Treatment

With drastic improvements in the treatment of childhood cancers, more attention has been given to the quality of life of cancer survivors. Furthermore, cancer and its effect on fertility have become a major focus of extensive research over the past decade. This amplified interest is due to the increased incidence of cancer among

L.I. Lipshultz, M.D. (✉)
Lester and Sue Smith Chair in Reproductive Medicine, Baylor College of Medicine, Houston, TX, USA
e-mail: larryl@bcm.edu

E. Seli and A. Agarwal (eds.), *Fertility Preservation in Males: Emerging Technologies and Clinical Applications*, DOI 10.1007/978-1-4614-5620-9_5, © Springer Science+Business Media New York 2012

the reproductive age group as well as improvements in long-term survival rates [2]. Cancer treatment encompasses three main treatment modalities: surgery, radiation therapy, and chemotherapy.

Testicular cancer unlike other malignancies exert a direct effect on spermatogenesis through the destruction of surrounding tissue, altering local blood flow which increases intrascrotal temperature not only in the affected testicle but the contralateral one as well [3]. These tumors can also produce hormones (human chorionic gonadotropin and alpha feto-protein) and effect spermatogenesis. Radical orchiectomy remains the mainstay of treatment for testicular tumors with partial orchiectomies recently becoming more popular as a method of preserving hormonal and sperm function in select patients [4]. Petersen et al. reported a 50% decrease in the sperm concentration during the first few months after radical orchiectomy, and found that 10% of patients with sperm prior to surgery will become azoospermic [5].

As an adjunct to radical orchiectomy, a subset of patients may proceed to a retroperitoneal lymph node dissection, which potentially causes further impairment in fertility. With this node dissection, despite template modifications and intentional nerve sparing techniques, damage to the sympathetic ganglia responsible for emission and ejaculation can still occur. For these patients who are anejaculatory after surgery, surgical sperm harvest and percutaneous extraction for assisted reproduction is an option; however, less invasive options include electroejaculation to stimulate antegrade ejaculation, post-ejaculate urinalysis as well as α-adrenergic therapy can be used [6, 7].

The effect of radiation therapy is entirely dependent on the dosage, treatment field and fractionation schedule [8]. Radiation doses start to adversely affect spermatogenesis at 0.1–1.2 Gy, with irreversible damage at doses >4 Gy [9]. However, higher doses (>20 Gy) may also affect the Leydig cells resulting in reduced testosterone and raised serum gonadotropins [10]. Although the recovery of spermatogenesis can take up to 9 years after treatment, updated radiation technologies with more efficient dosages and mandated gonadal shielding has allowed for a faster recovery of sperm production. Howell and Shalet reported complete recovery within 12 months after treatment with doses of ≤1 Gy, after 30 months with doses of 2–3 Gy, and >5 years with doses >4 Gy [11]. In addition, irradiation may increase sperm DNA damage that persists for up to 2 years after treatment, affecting fertility rates even after recovery of spermatogenesis [9].

Chemotherapeutic medications have a negative effect on fertility, although some agents are more gonadotoxic than others. Similarly to radiation therapy, the toxic effect on spermatogenesis is dependent on dosage, frequency, and pretreatment semen quality; however, recovery is also contingent upon the mechanism of action of each agent as well as the synergistic effect of multiple agent regimens [12]. Table 5.1 reviews the effect of standard regimens on gonadal function [12, 13]. The most gonadotoxic agents are the nitrogen mustard derivatives, such as busulphan, and alkylating agents, such as cyclophosphamide and isophosphamide which result in permanent azoospermia in 80–90% of cases [14]. Treatment of Hodgkin's disease with the nonalkylating combination of adriamycin, bleomycin, vinblastine, and

Table 5.1 Cytotoxic agents

Group	Agent	Mechanism of action	Gonadotoxicity
Alkylating agents	Cyclophosphamide ifosfamide Carmustine busulfan	Adds alkyl groups to DNA, altering structure and function	Prolonged azoospermia
Antibiotics	Actinomycin-D bleomycin	Binds to DNA inhibiting RNA synthesis single and double strand breaks in DNA	Possible azoo-spermia Temporary reduction in sperm count
	Doxorubicin	Inhibits progression of topoisomerase II dependent DNA	Temporary reduction in sperm count; may cause prolonged azoospermia when used with other agents
Antimetabolites	Fluorouracil 6-Mercaptopurine Thioguanine Methotrexate	Pyrimidine analog Purine analog Antifolate	Temporary reduction in sperm count with standard regimens, although additive effects are possible
Plant derivatives	Vincristine Vinblastine Etoposide	Inhibits formation of microtubules Inhibits topoisomerase II	Temporary reduction in sperm count
Platinum analogs	Carboplatin cisplatin	Forms DNA complex and interferes with repair mechanisms	Prolonged azoospermia
Miscellaneous	Interferon-α Prednisone	Stimulates macrophages and NK Cells suppresses immune response	No known affects on spermatogenesis

Data from Giwercman et al. [11] and Lee et al. [12]

dacarbazine revolutionized chemotherapy regimens, with a sperm recovery rate of ~90% within 1–5 years posttreatment while still maintaining therapeutic efficacy [15]. Germ cell tumors are typically treated with the regimen of bleomycin, etoposide and cisplatin (BEP), with patients demonstrating recovery rates back to baseline of 50% within 2 years and 80% within 8 years of treatment [11].

Numerous groups have examined the incidence of sperm aneuploidy after administration of chemotherapy by sperm fluorescent in situ hybridization (FISH). Multiple studies examining the effect of BEP on patients treated for testicular cancer demonstrated increased sperm aneuploidy that typically returned to baseline within 2 years after treatment [16, 17]. Patients treated for Hodgkin's lymphoma receiving either adriamycin, bleomycin, vincristine, and dacarbazine or novantrone, oncovin, velban, and prednisone had an increased incidence of sperm disomy and diploidy after therapy that typically returned to baseline months after treatment [18]. Advances in chemotherapy regimens continue to strive for treatment efficacy while minimizing side effects, particularly gonadal toxicity.

Effects of Chemical Exposures

Male reproductive health in the general population has garnered increasing attention due to reports of a time-related decline of semen quality, increased urogenital anomalies, and increased rates of testicular cancer [19–21]. The possibility that environmental chemicals may be partially responsible for these adverse outcomes has been a real concern since the discovery in 1977 of the severe spermatotoxic effect of dibromochloropropane (DBCP), a nematocide widely used in the USA and Central America, among workers at a chemical plant [22]. Furthermore, Anway et al. demonstrated that there are transgenerational effects of chemicals, where exposure of the parental gametes to chemicals may confer an increased risk of altered semen quality in the offspring [23]. Although the list of potential environmental gonadotoxicants is exhaustive, we focus on phthalates, pesticides, bisphenol A (BPA), and glycol ethers.

Phthalates

Phthalates are among the most widely used manmade chemicals released into the environment over the last several decades. They are primarily used as plasticizers in the manufacture of flexible vinyl, which is found in medical devices, toys, floor and wall coverings, personal care products, and food packaging. The ubiquitous use of phthalates results in exposure to these compounds mainly through diet, medical devices, and consumer products. Humans quickly metabolize phthalates to their respective di and monoesters; however, their metabolites have been found in the urine of more than 95% of men and women that have been investigated [24]. There have been several epidemiological studies addressing the male reproductive toxicity of phthalates. Duty et al. [25] observed an inverse dose–response relationship between phthalate metabolites and sperm concentration and motility. Another group also demonstrated an association between phthalate metabolite exposure and increased levels of DNA damage in sperm [26].

Pesticides

Pesticides form a large group of heterogeneous chemicals which are used to control insect, weed, rodent, and fungi populations. There are several epidemiological studies from across the globe on men exposed to contemporary-use pesticides during agricultural work. In a recent Japanese study, pesticide sprayers exposed primarily to organophosphates showed spraying season dependent reductions in motile sperm compared with unexposed controls [27]. One of the first studies demonstrating a link between specific biomarkers of environmental exposure to pesticides and biomarkers of male reproduction was performed in the USA [28]. Men demonstrated an inverse dose–response relationship between herbicide exposure (DBCP) and sperm concentration, motility, and morphology.

Men exposed to pesticides have also been examined for sperm aneuploidy with FISH. Recio et al. [29] demonstrated that male agricultural workers exposed to organophosphate pesticides with increased serum concentrations had an increased frequency of sperm aneuploidy. However, a study conducted by Smith et al. [30] failed to demonstrate an association between exposure to miscellaneous pesticides and sperm aneuploidy in men with low levels of exposure. There are numerous indications that certain pesticides may impair semen quality; however, the direct affect noted with pesticides, such as DBCP, fortunately are no longer as evident with contemporary-use pesticides.

Bisphenol A

BPA is a chemical with weak estrogenic activity and widely used in polycarbonate plastic products, epoxy resins, sealants, and the lining of food cans. BPA has been reported to have estrogenic effects in animal models, such as decreased sperm production, stimulation of prolactin release, and promotion of cell proliferation in a breast cancer cell line [31, 32]. BPA has been shown to leech from products due to incomplete polymerization and degradation by exposure to high temperatures [33]. BPA levels have been measured in the serum and urine of adults, breast milk, fetal plasma, as well as amniotic fluid [34–36]. Levels of BPA in the serum have also been correlated with circulating androgen levels, and found to be elevated in women with polycystic ovarian syndrome [37].

Women carrying fetuses with karyotype abnormalities, furthermore, were found to have elevated maternal serum BPA levels [38] while BPA levels were found to be three times higher in 45 women with three or more consecutive first trimester miscarriages [39]. The harmful potential of BPA has made such an indelible impact that BPA has been electively removed from the manufacturing of baby bottles sold in the USA by the six largest manufacturers.

Glycol Ethers

Glycol ethers are a family of alkyl derivatives of ethylene and propylene glycol. They are most commonly found in industrial uses, such as semiconductors, photographic film, paints, textile dyes, varnishes, printing inks, and hydraulic fluids. The oxidative metabolites of glycol ethers, particularly methoxyacetic acid (MAA), have been shown to be teratogenic and embryotoxic. In animal studies, MAA has altered gene expression and protein kinase activity, increased oxidative stress, and enhanced transcription of nuclear receptors for progesterone, estrogen, and androgens [40, 41]. The deleterious effects of glycol ether exposure have been reported for years. Over two decades ago, shipyard workers exposed to glycol ethers in paint fumes were found to have increased incidences of oligospermia [42]. More recently, men attending fertility clinics were found to have low motile sperm counts directly related to glycol ether exposure [43].

Lifestyle Factors

Lifestyle exposures, including ethanol, tobacco, recreational drug use, and caffeine have all been studied in relationship to male reproductive health and the focus has more recently been directed at semen quality and subsequent fertility. These substances have a substantial impact because of their frequent use and global exposure.

Ethanol is one of the most abused substances worldwide. It has been shown to have deleterious effects at all levels of the male reproductive system. Ethanol interferes with feedback mechanisms of the hypothalamus-pituitary-gonadal (HPG) axis resulting in impairment of production and secretion of adequate levels of luteinizing hormone and follicle-stimulating hormone [44]. Additionally, ethanol has been associated with significant decreases in serum testosterone levels, increased rates of plasma testosterone clearance, increased levels of plasma steroid hormone-binding globulin (SHBG) and increased levels of estrogen [45]. These effects are dose dependent and are potentially irreversible in chronic abusers, particularly increased levels of SHBG [46]. Ethanol has demonstrated direct effects on spermatogenesis which have manifested as a reduction in seminiferous tubular diameter with subsequent damage to the germinal epithelium, reduction in sperm count, reduction in progressive motility, reduction in sperm with normal morphology, as well as an increase in reactive oxygen species (ROS) [47, 48]. While few studies have addressed the effects of paternal ethanol exposure on offspring, animal studies have demonstrated significantly reduced fecundity in peripubertal ethanol-fed fathers [49].

Similarly, one-fifth of Americans *smoke* and nearly 50% of nonsmokers are exposed to secondhand smoke. Shockingly, 25% of Americans between the ages of 18 and 24 smoke and the number continues to rise [50]. Smoking is a lifestyle hazard for both the active and passive smoker. While much is known of the carcinogenic properties of tobacco and its resultant effects on organs, such as lung and bladder, the impact on fertility still remains less defined. Numerous studies, however, have well established the fact that toxins in cigarette smoke reach the male reproductive system and their effects are mainly due to their direct interaction with the components of seminal fluid [51]. This interaction has led to a greater presence of ROS, increased leukocytes and round cells, as well as a higher frequency of DNA fragmentation in comparison to nonsmokers [52–54]. In smokers, either male or female, there was noted to be a significant delay of over 6 months in natural conception in comparison to nonsmokers [55]. Nicotine and its metabolites are present in the spermatozoa of smokers, and furthermore, these toxicants are found in embryos resulting from in vitro fertilization (IVF) cycles with male smokers [56].

Marijuana is the most common illicit drug in the USA, and its abuse and dependence by users are steadily increasing [57]. It has been well established that the active components of marijuana have a negative impact on sperm function and fertilizing ability since it leads to oligospermia, an association with pyospermia as well as the development of gynecomastia [58, 59]. This is incredibly important, given the prevalence of marijuana use particularly in the young, as well as the more recent development of legalization of cannabinoid use for the treatment of specific medical conditions in certain countries. The effect of marijuana use on male fertility has to be taken into careful account.

Multiple researchers have examined the effect of *smoking* on sperm aneuploidy. Despite inconsistent results, combined data from each study demonstrated a significant increase in sex chromosome and autosome aneuploidy [60]. While controlling for smoking and alcohol consumption, Robbins et al. [61] found an increase in sex chromosome disomy and diploidy with high levels of *caffeine* intake while Rubes et al. [62] did not find an increase in either when compared to controls. Although numerous articles demonstrate deleterious effects of *alcohol* on semen parameters, only one article examined alcohol consumption by sperm FISH and found an association between increased alcohol consumption and sex chromosome disomy [61]. The potential confounding effects of smoking, caffeine, and alcohol intake in each of these articles makes drawing any broad conclusions difficult without rigorously controlled prospective studies.

Medications

The usage of certain drugs, and specific drug classes, may contribute to infertility. Antihypertensives are a class of drugs that has well-documented side effects, including reduced sperm quality. Spironolactone, a potassium-sparing diuretic, has proven antiandrogen activity and shown to reduce semen quality in men on chronic therapy [63]. Calcium channel blockers have also demonstrated a reversible functional deficit in sperm quality and a significant interference with fertilizing capability, although there was no change in sperm production noted [64]. Antibiotics are another class of drugs that have notoriously had significant side effects. Historically, nitrofurantoin given chronically at high doses (10 mg/kg/day) to young males has been reported to result in early maturation arrest; however, this has not been seen in clinical practice [65]. In vitro studies have also highlighted the potential to decrease fertility with the use of erythromycin, tetracycline, and gentamicin [66]. Again, there does not appear to be clinically supported relevance.

Other medications that have negatively affected fertility include sulfasalazine and colchicine, leading to reversible defects in sperm concentration and motility [67, 68].

Baumgartner et al. [69] described two men with a significant increase in sperm aneuploidy who had chronically ingested diazepam for greater than 6 months. Another group reported that finasteride was associated with increased sperm diploidy and sex chromosome disomy, which did not improve 1 year after cessation of the drug [70].

Male Evaluation

When performing the male evaluation for infertility, it is important to simulate the work up of any medical patient. A thorough history should be performed, addressing the amount of time in which the couple has attempted conception, previous pregnancies, intercourse timing, use of lubrication and erectile function. A query of medical conditions and surgical conditions should also be noted. Special attention should be paid to a developmental history, recent febrile illness, a history of viral orchitis, broncho-pulmonary illnesses, undescended testes, childhood cancers and treatments, genitourinary tract infections, and congenital or genetic abnormalities. Scrotal, prostatic, spinal, inguinal, and retroperitoneal surgery should be highlighted along with a family history of both infertility and genetic abnormalities. A medication history including the use of anabolic steroids, either past or present, should be elicited. Finally, occupational and environmental exposures, such as contact with toxins, chemicals, radiation, ethanol intake, and smoking should be noted.

The physical exam should include a general description of the patient, absence or presence of facial and pubic hair, gynecomastia and skeletal structure. The examination should include a complete penile exam, which includes the length of the phallus and position of the meatus. The scrotum should be examined for the size and consistency of the testes, presence and consistency of the epididymis, vas deferens, and the presence of a varicocele(s). The digital rectal exam should also begin with an examination of the position of the anus, sphincter tone, and a thorough prostate exam evaluating its size, consistency, and the presence of midline cysts or dilation of the seminal vesicles.

Spermatogenesis is a complex step-wise process, which relies on an adequate and functional signaling pathway. A testosterone and FSH should be obtained to evaluate the HPG axis. If abnormal values are discovered, then more extensive blood tests should be implemented which may include estradiol (if the patient is obese), LH, prolactin, TSH, SHBG, and cortisol. Although endocrinopathies account for less than 3% of infertility cases, a thorough hormonal evaluation is essential when appropriate.

Male infertility can be divided into two main components: inadequate sperm production or insufficient sperm delivery. The patient should have at least two semen samples analyzed and referenced to well-established World Health Organization (WHO) criteria. In the absence of sperm, the specimen should be centrifuged, the pellet evaluated, and a quantitative fructose performed. The physician may also adjunctively order one of many additional studies, i.e., strict morphology, DNA integrity testing, and ROS quantitative, if suggested by abnormalities in the semen analysis (SA).

Table 5.2 WHO semen parameter reference values

Semen parameter	Range
Volume	≥1.5 mL
Sperm concentration	≥15 M/mL
Total sperm count	≥39 M
Motility	≥40% total motility
	≥28% progressive (a+b) motility
Morphology	≥4% by strict criteria
Vitality	≥58% sperm viable
WBC	<1 M/mL

From Cooper et al. [74], with permission of Oxford University Press

Integrating the history, physical examination, SA, and endocrine profile are steps toward establishing the diagnosis of the infertile male. By utilizing this systematic testing, the patient can be further characterized as primary, secondary, or tertiary testicular failure; obstructed vs. nonobstructed; or having abnormalities of sperm quantity and quality. During this process, an effort is made to identify not only the underlying conditions and harmful exposures leading to male infertility, but also to determine the potential effectiveness of specific vs. empiric therapies, as well as likely candidacy for assisted reproductive techniques (ART).

Semen Analysis

The most important aspect of the initial laboratory evaluation is the SA. It is recommended that at least two semen analyses be obtained with similar abstinence periods. Wide variations in semen parameters between specimens from the same patient are possible. Reports indicate that sperm density increases 25% per day for the first 4 days of abstinence prior to an SA, but other parameters such as motility and morphology remain stable [71].

A word of caution when classifying a man as being fertile, subfertile, or sterile based solely on semen parameters. While it may be true that a man with azoospermia is considered sterile, there is a wide range of overlap between semen parameters of fertile compared to infertile men [72]. The WHO has published reference values for semen, including volume, pH, sperm concentration, total sperm count, motility, morphology, and others (Table 5.2) [73, 74]. Medical professionals and reproductive specialists throughout the world have increasingly utilizing these reference values in the evaluation of the infertile couple.

Semen samples should be obtained correctly by either masturbation or ejaculation into a specialized nonlatex spermatocidal-free collection condom. Collection of the sample can be performed in the office, or it may be collected at home and brought in for processing. In the case of the latter, the sample must be stored at room temperature and delivered within an hour of collection.

Once received, the ejaculated specimen is allowed to liquefy for 20–30 min. Gentle pipetting of the sample can aid in situations when there is incomplete liquefaction. This same technique is used to homogenize the sample prior to performing sperm counts. Alternately, one may use mechanical agitation with a vortex for several seconds. Viscosity is considered normal if semen exits the pipette drop by drop, and abnormal if it strands.

Microscopic Examination

Microscopic survey of semen must be performed prior to actual sperm counting. This is done to identify round cells, debris, and bacteria. Contamination of the specimen during collection as well as urethral or prostatic bacteria can result in the presence of bacteria in the semen. Significant infection is seen if concentration exceeds 1,000 bacteria per millimeter [75]. The presence of round cells in the semen can represent either immature germ cells or leukocytes [76]. Of note, up to 20% of infertile men have leukocytes present in the semen and of these only 20% are associated with significant bacterial colonization [77].

Sperm Counting

The preferred method for determining sperm density ("counts") is by using commercially available Neubauer hemocytometer counting chamber slides. Older methods required dilution of the specimen, which confounded a technique with an already inherently high technical error. Despite using counting chambers there still exists a significant variation in results between each type of chamber, whether computer based or manually determined [78, 79]. In general, one should count at least 200 sperm for an adequate determination of sperm concentration. When using microscopes that integrate a grid for counting, all sperm is counted within the grid and the concentration is calculated as a function of a coefficient specific to each chamber [80].

Sperm Motility

Percent motility has been shown to be the most important parameter when correlating semen samples to pregnancy outcomes [81, 82]. It is also the most difficult part of the manual evaluation of the SA and introduces a significant amount of subjectivity across technicians. One method involves a subjective estimation of motility from surveying several fields by the technician and averaging those estimates to produce a motility percentage.

A more objective and accepted method is to count motile and nonmotile (NM) sperm in each grid and average the percentages to produce a motility value. This can be done with a manual hematology cell counter. Drawbacks of this method are that it is inherently time consuming and that overestimation of motility can occur. This is due to the fact that as the technician moves along the grid, very motile sperm may progress to other areas of the grid and, theoretically, be double counted.

Methods to neutralize the inherent difficulty of counting motile sperm have been proposed [80]. Once the semen sample is homogenized, an aliquot is prepared in parallel for counting. In one aliquot only the NM sperm are counted. In the other, an immobilizing agent is used, usually water, after which all sperms (T) are counted. Motility can be calculated by using the formula T-NM.

Strict Sperm Morphology

The traditional evaluation of sperm morphology classifies sperm as normal if they do not fit into one of several defined categories, although the majority of sperm in an ejaculate are neither uniform nor symmetrical, displaying large variation in shape and size. A common classification scheme designates sperm as normal (oval), amorphous, tapered, duplicated, and immature. However, more strict criteria have been developed to identify "normal" spermatozoa [83, 84]. Kruger et al. reported that using his strict criteria, patients with less than 4% normal forms had a fertilization rate of 7.6% of oocytes in comparison to the usual >50% in patients with 4–14% normal forms [83].

For determining strict criteria, a minimum of 100–200 spermatozoa are counted on a stained slide using 100× oil-immersion magnification. An eyepiece reticule is initially recommended for measuring the sperm head and tail size (length and width), although this may not be necessary with an experienced andrology technician. Using the strict morphological criteria, a normal spermatozoon is characterized by a smooth oval head, 4.0–6.0 μm in length and 2.4–3.5 μm in width. The acrosome must be well defined, covering 40–70% of the sperm head. There cannot be any mid piece or tail defects. Finally, there should be no cytoplasmic droplets greater than half the size of the sperm head. Utilizing the strict criteria method, a spermatozoon that might be considered "borderline" would be classified as abnormal [84].

Although strict morphology criteria have been widely accepted, its clinical usefulness remains an area of controversy. Sperm morphology was not a reliable predictor in selecting sperm without chromosomal aberrations [85] while normal morphology was not a good indicator of genetically normal sperm [86]. Guzick et al. reported the existence of significant overlap between fertile and infertile men; and sperm morphology, concentration, or motility was not a powerful discriminator between the two groups [72].

While the establishment of high quality control standards in evaluating morphology may improve the clinical utility of morphology determination, couples should be counseled with caution to pursue ART based solely on an abnormal strict morphology.

Reactive Oxygen Species

Oxidative stress is a recognized etiology of male infertility [87]. ROS in the form of superoxide anions, hydrogen peroxide, and the hydroxyl-free radical are formed as a by-product of oxygen metabolism. The presence of excess ROS can cause oxidative damage to lipids, proteins, and DNA [88–90]. Abnormal ROS formation is found in up to 40% of infertile patients [91], with some reports suggesting an inverse relationship between seminal ROS levels and spontaneous pregnancy outcomes of infertile couples [92]. Many studies have attempted to define the relationship between seminal ROS and IVF [93, 94] but have met with conflicting results. Nevertheless, a growing body of knowledge on ROS and fertility makes testing for oxidants in the semen an important part of the infertile male evaluation.

There are various methods to detect seminal ROS, including chemiluminescence, nitroblue tetrazolium test, cytochrome C reduction test, flow cytometry, electron spin resonance, and xylenol orange-based assay. We discuss the chemiluminescence method, as this is the most commonly used.

The chemiluminescence assay utilizes a luminometer to measure chemical reactions between ROS found in human semen and a chemiluminescent probe, such as luminol or lucigenin. Luminol is an uncharged particle that is cell membrane permeable and therefore can react extracellularly and intracellularly with hydrogen peroxide, hydroxyl anions, and superoxide anions. In contrast, lucigenin is a positively charged particle that is membrane impermeable and reacts with superoxide anions in the extracellular space [95].

It is important to remember that the majority of the assays for chemiluminescence measure total oxidative stress and do not generally distinguish between ROS produced by leukocytes and that produced by spermatozoa. Moreover, there are other factors that can alter these assay results and must be considered.

Leukocyte contamination in the semen has been shown to impact negatively on fertility [77]. Leukocytes have been shown to be responsible for a significant proportion of ROS activity in the semen [96, 97]; therefore, these assays should be coupled with selective leukocyte removal strategies if pyospermia is present [98]. Not doing so would lead to a falsely elevated ROS value. Other factors that may spuriously increase ROS results include repeated centrifugation of the sample [99] and the use of certain oxidase-containing buffers for sample preparation [100].

In contrast, prolonged time from preparation to analysis of the sample can artificially decrease the ROS identifiable in the semen [101]. For this reason, it is recommended that testing be performed within an hour of sperm preparation [102]. Finally, poor liquefaction of the sample can interfere with the normal oxidative process resulting in a falsely lower ROS value [100].

DNA Damage

High levels of sperm DNA damage can negatively impact reproduction. Sperm samples from infertile men have been shown to have significantly more DNA damage than their fertile counterparts [103–105]. In addition, DNA damage has been correlated with poor outcomes from intrauterine inseminations [106]. Multiple reports have also implicated DNA damage with poor IVF results [107, 108], although newer studies have not corroborated these findings [109–111]. Finally, some evidence suggests that DNA fragmentation can be a cause of early embryo death, poor embryo progression and poor implantation [112, 113]. In general, damage to sperm DNA occurs during intratesticular development as well as during the maturation and transport process that takes place outside the testis.

It is known that spermatogenesis is controlled by selective apoptosis. Abnormal sperm is tagged for apoptosis in the same manner that all other cells are marked for programmed cell death. Evidence suggests that a malfunction in this process allows sperm with DNA damage to be transported in the ejaculate, a process referred as abortive apoptosis [114, 115]. Protamine deficiency has been identified as another primary testicular cause of sperm DNA damage, and this deficiency is frequently seen in infertile men compared to fertile counterparts [116]. In addition, certain polymorphisms in the protamine gene have been implicated in male infertility and sperm DNA damage [117].

Excessive ROS in the ejaculate correlates with increasing sperm DNA damage [88, 97, 118]. Fortunately, there is evidence suggesting that reduction in ROS levels with antioxidant therapies can decrease sperm DNA damage [103, 119]. Studies have also implicated clinically significant varicoceles as a cause of sperm DNA damage [120]. Recent reports suggest improvement if DNA damage following microsurgical varicocelectomy [121–123]. Other external factors implicated in sperm DNA damage include smoking and genital tract inflammation [124, 125].

Since assisted reproductive technologies are now commonly used to circumvent virtually all types of male infertility, it is important to understand the differences between ejaculated, epididymal, and testicular sperm and their respective levels of DNA damage. O'Connell and colleagues found that testicular sperm had fewer DNA mutations and fragmentations when compared to epididymal sperm in preparation for IVF/intracytoplasmic sperm injection (ICSI) [126]. When comparing testicular sperm to ejaculated sperm, Greco and colleagues found that there was significantly lower DNA fragmentation in the testicular sperm. In addition, they reported improved pregnancy rates using testicular sperm compared to ejaculated sperm [127]. If IVF/ICSI is to be performed using sperm with high DNA damage consideration should be given to testicular sperm extraction (TESE) only after less invasive treatments for known causes of DNA damage have failed.

Common Genetic Abnormalities in the Male

Over the last several decades, advances in the field of infertility have enabled physicians to better evaluate the genetic causes of male factor infertility. It is important that patients, who are found to have genetic abnormalities or those who are suspected, be counseled on the risk to the fetus by either a genetic counselor or an experienced and knowledgeable physician. Proper risk assessment should be discussed with the patient as well as proposals for pre- and postconception genetic testing. Limitations in currently available genetic testing are highlighted in cases, where there is a high suspicion of a genetic cause, but current testing yields negative results, underscoring the importance of education and counseling for these couples.

Males who are found to have impaired semen quality should be referred to a male infertility specialist prior to the couple undergoing ART. The objective of the evaluation is to identify the etiology the infertility, realizing, however, that a significant percentage of these cases will remain idiopathic. Causes of abnormal semen can be "genetic," hormonal, mechanical or the result of concurrent disease or cancer [128, 129]. The physician should next attempt to either enhance the patient's sperm concentration for natural conception, intrauterine insemination or if necessary, proceed with extraction of sperm for IVF/ICSI. The infertility specialist should also be aware of common genetic abnormalities and proper testing protocols for suspected patients.

In patients with genetic abnormalities undergoing ICSI, preimplantation genetic testing can be used in conjunction with IVF/ICSI. Preimplantation genetic diagnosis (PGD), which involves the removal and analysis of genetic material from the embryo, is indicated in patients that have a risk of transmitting genetic disorders or mutations to their offspring. PGD can also be used in combination with other prenatal diagnostic testing, such as chorionic villous sampling, amniocentesis, blood testing, and ultrasound. Risks should be thoroughly discussed, such as increased risk of chromosomal abnormalities, prenatal mortality, multiple gestations, low birth weight, and preterm delivery with IVF/ICSI as well as the possibility of misdiagnosis and subsequent fetal loss with current diagnostic techniques [130–133].

CFTR: Absence of the Vas Deferens

The vas deferens can be palpated easily in the majority of males during the physical exam. In cases in which there is unilateral or bilateral absence of the vas deferens further evaluation is warranted. These patients also have a higher incidence of an abnormal or missing renal unit. Furthermore, 25% with unilateral vasal agenesis (CUAVD) and 10% with bilateral (CBAVD) will have abnormal seminal vesicles found on ultrasound [134]. Mutations in the cystic fibrosis transmembrane conductance regulator (CFTR) gene are responsible for many cases of CBAVD and CUAVD. Approximately 2/3 of men with CBAVD will have a mutation in the CFTR gene, and this abnormality effects 1–2% of men presenting with infertility [135]. However,

10–40% of men that are found to have CBAVD do not have subsequent mutations with routine CFTR gene testing [136]. In cases where there are positive physical findings and a negative test, the female partner should be tested. The majority of males with either CBAVD or CUAVD have normal semen parameters. Sperm can be retrieved from these patients via microscopic epididymal sperm aspiration (MESA), TESE or percutaneous testicular sperm aspiration (PESA), and ART can be used.

Klinefelter's Syndrome

Identifiable chromosomal abnormalities account for approximately 5% of male factor infertility [137]. In patients that are found to be azoospermic, the rate increases threefold [138]. Aneuploidy occurs in instances in which there are superfluous or insufficient numbers of chromosomes. The most common syndrome associated with aneuploidy is Klinefelter's disease, which occurs in 1: 500 births and is found in 15% of males with infertility. The most common chromosomal arrangement in Klinefelter's disease is nonmosaic 47,XXY or mosaic 47,XXY/46X. These patients commonly have some degree of spermatogenic dysfunction that ranges from severe oligospermia to azoospermia. Testicular sperm retrieval techniques can be utilized for ICSI. Chromosomal abnormalities can be identified by a karyotype of peripheral leukocytes. Klinefelter's patients should be counseled on the increased risk of autosomal and sex chromosomal abnormalities in children conceived through IVC/ICSI [139].

Robertsonian Translocations

Robertsonian translocations are the most common form of balanced chromosomal rearrangement found in humans. The disorganization of the chromosomes can cause a loss of genetic material and/or abnormal signaling. The most common chromosomes involved are chromosomes 13, 14, 15, 21, 22, and occur in approximately 1 in 1,000 births [140]. Carriers can appear phenotypically normal; however, these patients are frequently present with oligospermia or azoospermia. These patients are also at higher risk for repeat miscarriages, failed IVF cycles and have a higher risk for offspring to be affected with unbalanced Robertsonian translocations, such as trisomy 21 and 13 [141]. Individuals with significant risk factors are recommended to undergo sperm FISH and PGD.

Y Chromosome Deletion

Y chromosome deletions occur in 5–10% of infertile males presenting with nonobstructive sperm deficiencies. The Y chromosome contains vital components needed for male differentiation and sperm function. The azoospermia factor region

(AZF) on the long arm of the Y chromosome (Yq) is responsible for sperm development. The AZF region is subdivided by location into AZFa, AZFb, and AZFc, which correspond to proximal, middle, and distal portions of the chromosome [142]. Deletions in these locations are responsible for varying degrees of spermatogenic dysfunction. Entire microdeletions of AZFa or AZFb regions of the Y chromosome portend an exceptionally poor prognosis in sperm retrieval, such that microscopic sperm extraction is predictably negative [143].

Whereas microdeletions in AZFc alone can result in a variety of phenotypes which include oligospermia and azoospermia, depending on the severity of the deletion [143]. Deletions in the Y(q) are too small to be detected with a karyotype and thus are termed microdeletions. Identification using polymerase chain reaction techniques to analyze sequence tagged sites is currently used. Indications for testing AZF microdeletions are sperm concentrations less than 5 million/mL. Importantly, male offspring of patients with Y microdeletions will inherit the abnormal gene, rendering them likely to be infertile. Thus, ICSI with PGD should be discussed.

Significance of Assisted Reproduction

Over the past three decades, the treatment of infertile couples has been revolutionized by ART, including the development of ICSI and PGD. However, unlike most therapeutic procedures used in medicine today, ARTs have not undergone the rigorous safety testing prior to clinical use. With these new technologies, we are able to overcome many types of sterility and bypass nature's protective mechanisms that were developed in evolution to prevent fertilization by defective or deficient sperm. Subsequently, large numbers of couples undergo fertility treatments without a complete understanding of their infertility or the potential long-term risks for their children.

Historically, the treatment of male factor infertility with IVF was largely unsuccessful; yet the development of ICSI revolutionized the treatment of severe male factor infertility, affording many men the opportunity to father their own biologic children. However, evidence exists with regard to risks to the mother and offspring that are significantly increased with ART, including multiple gestation, preterm delivery (even in singleton pregnancies), and congenital anomalies in offspring [144–146]. Distinctive issues arise with regard to understanding the clinical implications of these data. Although most IVF pregnancies proceed uneventfully, studies consistently identify an increased risk of problems in IVF/ICSI pregnancies and deliveries [147].

Several large studies (Table 5.3) have shown a higher risk of genitourinary, cardiovascular, musculoskeletal, and gastrointestinal defects in children conceived by IVF and IVF/ICSI [148–157]. While current data suggest that IVF/ICSI is safe, it is still important to understand the nature of the risks and establish whether they are directly related to the technology itself or as a result of genetic defects from the parents. While controversial, there are further studies suggesting

Table 5.3 Association between congenital malformations and conception by assisted reproductive technologies

Study	Years studied	Study type and sample size	Findings
Hansen et al. [84], Australia	1993–1997	Registry, controlled: 1,138 ART children (301 IVF/ICSI, 837 IVF, 4,000 SC)	IVF and ICSI: ↑ likelihood of birth defects even after correction
Zhu et al. [85], Denmark	1997–2003	Registry/questionnaire, controlled: 64,405 children (SC: 50,897 singletons and 1,366 twins from fertile couples, 5,764 singletons and 100 twins from subfertile couples; ART: 4,588 singletons and 1,690 twins)	ART and subfertile spontaneous conceptions: ↑ congenital malformation with increasing delay to conception
Koivurova et al. [86], Finland	1990–1995	Registry, controlled: 304 IVF, 569 SC	IVF: ↑ adverse pregnancy outcomes before correction for multiplicity and cardiac malformations regardless of multiplicity
Ludwig and Kalainic [87], Germany	1998–2002	Registry, controlled: 3,372 IVF/ICSI, 30,940 SC (Mainz Birth Registry)	IVF/ICSI: ↑ risk of all adverse outcomes; RR 1.25 (95% CI 1.11–1.40)
Anthony et al. [88], The Netherlands	1995–1996	Registry, controlled: 4,224 ART, 314,605 SC	ART: ↑ cardiovascular malformation; differences minimal for any other congenital malformation
Ericson and Kallen [89], Sweden	1982–1997	Registry, controlled: 9,111 IVF, 1,690,577 SC	ICSI: ↑ risk of alimentary atresia, neural tube defects and hypospadias
Sutcliffe et al. [90], UK	1989–1994	Prospective, controlled: 91 ART (cryopreserved embryos), 83 SC	ART: differences not significant for any congenital malformation
Olson et al. [91], US	1989–2002	Registry, controlled: 1,462 ART, 8,422 SC	IVF: ↑ birth defects after correction for multiplicity
Lie et al. [92]	2005	Meta-analysis, controlled: 5,395 IVF/ICSI, 13,086 IVF	IVF/ICSI: no significant risk of any single congenital malformation
Hansen et al. [93]	2005	Meta-analysis, controlled: 28,638 ART	ART: ↑ any birth defect in all 25 studies or on analysis of seven well-designed studies

Adapted from Alukal and Lipshultz [161], reprinted with permission of Macmillan Publishers, Ltd
ART assisted reproductive technologies, *ICSI* intracytoplasmic sperm injection, *IVF* in vitro fertilization, *OR* odds ratio, *RR* relative risk, *SC* spontaneous conceptions

that developmental delay and impaired neurologic status occur in children conceived with IVF and IVF/ICSI [158, 159]. Furthermore, reporting on congenital abnormalities is inconsistent and at times inaccurate. Simpson and Lamb [160] outlined the limitations of IVF outcomes research and focused on the significant incidence of underreporting.

Given the increasing popularity of IVF/ICSI and the vast numbers of procedures performed each year throughout the world, continued research on the safety of IVF/ICSI and IVF itself is crucial. Patients can only truly give informed consent when they are properly educated as to all the associated risks. This information should be discussed with infertile couples prior to beginning the exhaustive journey, emotionally and financially, of the process of ART.

References

1. Thonneau P, Marchand S, Tallec A, et al. Incidence and main causes of infertility in a resident population (1,850,000) of three French regions (1988–1989). Hum Reprod. 1991;6:811.
2. Padron OF, Sharma RK, Thomas Jr AJ, et al. Effects of cancer on spermatozoa quality after cryopreservation: a 12-year experience. Fertil Steril. 1997;67:326.
3. Costabile RA, Spevak M. Cancer and male factor infertility. Oncology (Williston Park). 1998;12:557.
4. Shukla AR, Woodard C, Carr MC, et al. Experience with testis sparing surgery for testicular teratoma. J Urol. 2004;171:161.
5. Petersen PM, Skakkebaek NE, Rorth M, et al. Semen quality and reproductive hormones before and after orchiectomy in men with testicular cancer. J Urol. 1999;161:822.
6. Halstead LS, VerVoort S, Seager SW. Rectal probe electrostimulation in the treatment of anejaculatory spinal cord injured men. Paraplegia. 1987;25:120.
7. Kamischke A, Nieschlag E. Treatment of retrograde ejaculation and anejaculation. Hum Reprod Update. 1999;5:448.
8. Thomson AB, Critchley HO, Kelnar CJ, et al. Late reproductive sequelae following treatment of childhood cancer and options for fertility preservation. Best Pract Res Clin Endocrinol Metab. 2002;16:311.
9. Stahl O, Eberhard J, Jepson K, et al. Sperm DNA integrity in testicular cancer patients. Hum Reprod. 2006;21:3199.
10. Shalet SM, Tsatsoulis A, Whitehead E, et al. Vulnerability of the human Leydig cell to radiation damage is dependent upon age. J Endocrinol. 1989;120:161.
11. Howell SJ, Shalet SM. Spermatogenesis after cancer treatment: damage and recovery. J Natl Cancer Inst Monogr. 2005;34:12–7.
12. Giwercman A, Petersen PM. Cancer and male infertility. Baillieres Best Pract Res Clin Endocrinol Metab. 2000;14:453.
13. Lee SJ, Schover LR, Partridge AH, et al. American Society of Clinical Oncology recommendations on fertility preservation in cancer patients. J Clin Oncol. 2006;24:2917.
14. Pryzant RM, Meistrich ML, Wilson G, et al. Longterm reduction in sperm count after chemotherapy with and without radiation therapy for non-Hodgkin's lymphomas. J Clin Oncol. 1993;11:239.
15. Fossa SD, Magelssen H. Fertility and reproduction after chemotherapy of adult cancer patients: malignant lymphoma and testicular cancer. Ann Oncol. 2004;15 Suppl 4:iv259.
16. Martin RH, Ernst S, Rademaker A, et al. Analysis of sperm chromosome complements before, during, and after chemotherapy. Cancer Genet Cytogenet. 1999;108:133.

17. De Mas P, Daudin M, Vincent MC, et al. Increased aneuploidy in spermatozoa from testicular tumour patients after chemotherapy with cisplatin, etoposide and bleomycin. Hum Reprod. 2001;16:1204.
18. Tempest HG, Ko E, Chan P, et al. Sperm aneuploidy frequencies analysed before and after chemotherapy in testicular cancer and Hodgkin's lymphoma patients. Hum Reprod. 2008;23:251.
19. Carlsen E, Giwercman A, Keiding N, et al. Evidence for decreasing quality of semen during past 50 years. BMJ. 1992;305:609.
20. Paulozzi LJ. International trends in rates of hypospadias and cryptorchidism. Environ Health Perspect. 1999;107:297.
21. Huyghe E, Matsuda T, Thonneau P. Increasing incidence of testicular cancer worldwide: a review. J Urol. 2003;170:5.
22. Whorton D, Krauss RM, Marshall S, et al. Infertility in male pesticide workers. Lancet. 1977;2:1259.
23. Anway MD, Cupp AS, Uzumcu M, et al. Epigenetic transgenerational actions of endocrine disruptors and male fertility. Science. 2005;308:1466.
24. Wittassek M, Wiesmuller GA, Koch HM, et al. Internal phthalate exposure over the last two decades a retrospective human biomonitoring study. Int J Hyg Environ Health. 2007;210:319.
25. Duty SM, Silva MJ, Barr DB, et al. Phthalate exposure and human semen parameters. Epidemiology. 2003;14:269.
26. Hauser R, Meeker JD, Singh NP, et al. DNA damage in human sperm is related to urinary levels of phthalate monoester and oxidative metabolites. Hum Reprod. 2007;22:688.
27. Kamijima M, Hibi H, Gotoh M, et al. A survey of semen indices in insecticide sprayers. J Occup Health. 2004;46:109.
28. Swan SH, Kruse RL, Liu F, et al. Semen quality in relation to biomarkers of pesticide exposure. Environ Health Perspect. 2003;111:1478.
29. Recio R, Robbins WA, Borja-Aburto V, et al. Organophosphorous pesticide exposure increases the frequency of sperm sex null aneuploidy. Environ Health Perspect. 2001;109:1237.
30. Smith JL, Garry VF, Rademaker AW, et al. Human sperm aneuploidy after exposure to pesticides. Mol Reprod Dev. 2004;67:353.
31. Vom Saal FS, Cooke PS, Buchanan DL, et al. A physiologically based approach to the study of bisphenol A and other estrogenic chemicals on the size of reproductive organs, daily sperm production, and behavior. Toxicol Ind Health. 1998;14:239.
32. Steinmetz R, Brown NG, Allen DL, et al. The environmental estrogen bisphenol A stimulates prolactin release in vitro and in vivo. Endocrinology. 1997;138:1780.
33. Krishnan AV, Stathis P, Permuth SF, et al. Bisphenol-A: an estrogenic substance is released from polycarbonate flasks during autoclaving. Endocrinology. 1993;132:2279.
34. Calafat AM, Kuklenyik Z, Reidy JA, et al. Urinary concentrations of bisphenol A and 4-nonylphenol in a human reference population. Environ Health Perspect. 2005;113:391.
35. Ikezuki Y, Tsutsumi O, Takai Y, et al. Determination of bisphenol A concentrations in human biological fluids reveals significant early prenatal exposure. Hum Reprod. 2002;17:2839.
36. Sun Y, Irie M, Kishikawa N, et al. Determination of bisphenol A in human breast milk by HPLC with column-switching and fluorescence detection. Biomed Chromatogr. 2004;18:501.
37. Takeuchi T, Tsutsumi O, Ikezuki Y, et al. Positive relationship between androgen and the endocrine disruptor, bisphenol A, in normal women and women with ovarian dysfunction. Endocr J. 2004;51:165.
38. Yamada H, Furuta I, Kato EH, et al. Maternal serum and amniotic fluid bisphenol A concentrations in the early second trimester. Reprod Toxicol. 2002;16:735.
39. Sugiura-Ogasawara M, Ozaki Y, Sonta S, et al. Exposure to bisphenol A is associated with recurrent miscarriage. Hum Reprod. 2005;20:2325.
40. Bagchi G, Hurst CH, Waxman DJ. Interactions of methoxyacetic acid with androgen receptor. Toxicol Appl Pharmacol. 2009;238:101.

41. Miller RR, Carreon RE, Young JT, et al. Toxicity of methoxyacetic acid in rats. Fundam Appl Toxicol. 1982;2:158.
42. Welch LS, Schrader SM, Turner TW, et al. Effects of exposure to ethylene glycol ethers on shipyard painters: II. Male reproduction. Am J Ind Med. 1988;14:509.
43. Cherry N, Moore H, McNamee R, et al. Occupation and male infertility: glycol ethers and other exposures. Occup Environ Med. 2008;65:708.
44. Emanuele MA, Emanuele NV. Alcohol's effects on male reproduction. Alcohol Health Res World. 1998;22:195.
45. Muthusami KR, Chinnaswamy P. Effect of chronic alcoholism on male fertility hormones and semen quality. Fertil Steril. 2005;84:919.
46. Vicari E, Arancio A, Giuffrida V, et al. A case of reversible azoospermia following withdrawal from alcohol consumption. J Endocrinol Invest. 2002;25:473.
47. Villalta J, Ballesca JL, Nicolas JM, et al. Testicular function in asymptomatic chronic alcoholics: relation to ethanol intake. Alcohol Clin Exp Res. 1997;21:128.
48. Donnelly GP, McClure N, Kennedy MS, et al. Direct effect of alcohol on the motility and morphology of human spermatozoa. Andrologia. 1999;31:43.
49. Emanuele NV, LaPagli N, Steiner J, et al. Peripubertal paternal EtOH exposure. Endocrine. 2001;14:213.
50. Orleans CT. Increasing the demand for and use of effective smoking-cessation treatments reaping the full health benefits of tobacco-control science and policy gains in our lifetime. Am J Prev Med. 2007;33:S340.
51. Kulikauskas V, Blaustein D, Ablin RJ. Cigarette smoking and its possible effects on sperm. Fertil Steril. 1985;44:526.
52. Sepaniak S, Forges T, Fontaine B, et al. Negative impact of cigarette smoking on male fertility: from spermatozoa to the offspring. J Gynecol Obstet Biol Reprod (Paris). 2004;33:384.
53. Trummer H, Habermann H, Haas J, et al. The impact of cigarette smoking on human semen parameters and hormones. Hum Reprod. 2002;17:1554.
54. Saleh RA, Agarwal A, Sharma RK, et al. Effect of cigarette smoking on levels of seminal oxidative stress in infertile men: a prospective study. Fertil Steril. 2002;78:491.
55. Hull MG, North K, Taylor H, et al. Delayed conception and active and passive smoking. The Avon Longitudinal Study of Pregnancy and Childhood Study Team. Fertil Steril. 2000;74:725.
56. Zenzes MT, Puy LA, Bielecki R, et al. Detection of benzo[a]pyrene diol epoxide-DNA adducts in embryos from smoking couples: evidence for transmission by spermatozoa. Mol Hum Reprod. 1999;5:125.
57. Compton WM, Grant BF, Colliver JD, et al. Prevalence of marijuana use disorders in the United States: 19911992 and 2001–2002. JAMA. 2004;291:2114.
58. Close CE, Roberts PL, Berger RE. Cigarettes, alcohol and marijuana are related to pyospermia in infertile men. J Urol. 1990;144:900.
59. Rossato M, Pagano C, Vettor R. The cannabinoid system and male reproductive functions. J Neuroendocrinol. 2008;20 Suppl 1:90.
60. Robbins WA, Elashoff DA, Xun L, et al. Effect of lifestyle exposures on sperm aneuploidy. Cytogenet Genome Res. 2005;111:371.
61. Robbins WA, Vine MF, Truong KY, et al. Use of fluorescence in situ hybridization (FISH) to assess effects of smoking, caffeine, and alcohol on aneuploidy load in sperm of healthy men. Environ Mol Mutagen. 1997;30:175.
62. Rubes J, Lowe X, Moore II D, et al. Smoking cigarettes is associated with increased sperm disomy in teenage men. Fertil Steril. 1998;70:715.
63. Tidd MJ, Horth CE, Ramsay LE, et al. Endocrine effects of spironolactone in man. Clin Endocrinol (Oxf). 1978;9:389.
64. Benoff S, Cooper GW, Hurley I, et al. The effect of calcium ion channel blockers on sperm fertilization potential. Fertil Steril. 1994;62:606.
65. Nelson WO, Bunge RG. The effect of therapeutic dosages of nitrofurantoin (furadantin) upon spermatogenesis in man. J Urol. 1957;77:275.

66. Hargreaves CA, Rogers S, Hills F, et al. Effects of cotrimoxazole, erythromycin, amoxycillin, tetracycline and chloroquine on sperm function in vitro. Hum Reprod. 1998;13:1878.
67. Birnie GG, McLeod TI, Watkinson G. Incidence of sulphasalazine-induced male infertility. Gut. 1981;22:452.
68. Sarica K, Suzer O, Gurler A, et al. Urological evaluation of Behcet patients and the effect of colchicine on fertility. Eur Urol. 1995;27:39.
69. Baumgartner A, Schmid TE, Schuetz CG, et al. Detection of aneuploidy in rodent and human sperm by multicolor FISH after chronic exposure to diazepam. Mutat Res. 2001;490:11.
70. Collodel G, Scapigliati G, Moretti E. Spermatozoa and chronic treatment with finasteride: a TEM and FISH study. Arch Androl. 2007;53:229.
71. Carlsen E, Petersen JH, Andersson AM, et al. Effects of ejaculatory frequency and season on variations in semen quality. Fertil Steril. 2004;82:358.
72. Guzick DS, Overstreet JW, Factor-Litvak P, et al. Sperm morphology, motility, and concentration in fertile and infertile men. N Engl J Med. 2001;345:1388.
73. World Health Organization. WHO laboratory manual for the examination of human semen and sperm-cervical mucus interaction. 4th ed. Cambridge: World Health Organization, Published on behalf of the World Health Organization by Cambridge University Pres; 1999. p. 128.
74. Cooper TG, Noonan E, von Eckardstein S, et al. World Health Organization reference values for human semen characteristics. Hum Reprod Update. 2010;16:231.
75. Keck C, Gerber S, Chafer C, Clad A, et al. Seminal tract infections: impact on male fertility and treatment options. Hum Reprod Update. 1998;4:891.
76. Colpi GM, Lange A. Diagnostic usefulness of study of the round cells in seminal fluid. Acta Eur Fertil. 1984;15:265.
77. Wolff H. The biologic significance of white blood cells in semen. Fertil Steril. 1995;63:1143.
78. Prathalingam NS, Holt WW, Revell SG, et al. The precision and accuracy of six different methods to determine sperm concentration. J Androl. 2006;27:257.
79. Tomlinson M, Turner J, Powell G, et al. One-step disposable chambers for sperm concentration and motility assessment: how do they compare with the World Health Organization's recommended methods? Hum Reprod. 2001;16:121.
80. World-Health-Organization. WHO laboratory manual for the examination and processing of human semen. 5th ed. Geneva: World-Health-Organization; 2010.
81. Bostofte E, Bagger P, Michael A, et al. Fertility prognosis for infertile men: results of follow-up study of semen analysis in infertile men from two different populations evaluated by the Cox regression model. Fertil Steril. 1990;54:1100.
82. Mayaux MJ, Schwartz D, Czyglik F, et al. Conception rate according to semen characteristics in a series of 15 364 insemination cycles: results of a multivariate analysis. Andrologia. 1985;17:9.
83. Kruger TF, Menkveld R, Stander FS, et al. Sperm morphologic features as a prognostic factor in in vitro fertilization. Fertil Steril. 1986;46:1118.
84. Mortimer D, Menkveld R. Sperm morphology assessment historical perspectives and current opinions. J Androl. 2001;22:192.
85. Celik-Ozenci C, Jakab A, Kovacs T, et al. Sperm selection for ICSI: shape properties do not predict the absence or presence of numerical chromosomal aberrations. Hum Reprod. 2004;19:2052.
86. Ryu HM, Lin WW, Lamb DJ, et al. Increased chromosome X, Y, and 18 nondisjunction in sperm from infertile patients that were identified as normal by strict morphology: implication for intracytoplasmic sperm injection. Fertil Steril. 2001;76:879.
87. Aitken RJ. A free radical theory of male infertility. Reprod Fertil Dev. 1994;6:19.
88. Aitken RJ, Baker MA, Sawyer D. Oxidative stress in the male germ line and its role in the aetiology of male infertility and genetic disease. Reprod Biomed Online. 2003;7:65.
89. Griveau JF, Le Lannou D. Reactive oxygen species and human spermatozoa: physiology and pathology. Int J Androl. 1997;20:61.
90. Henkel R, Hajimohammad M, Stalf T, et al. Influence of deoxyribonucleic acid damage on fertilization and pregnancy. Fertil Steril. 2004;81:965.

91. Iwasaki A, Gagnon C. Formation of reactive oxygen species in spermatozoa of infertile patients. Fertil Steril. 1992;57:409.
92. Aitken RJ, Irvine DS, Wu FC. Prospective analysis of sperm-oocyte fusion and reactive oxygen species generation as criteria for the diagnosis of infertility. Am J Obstet Gynecol. 1991;164:542.
93. Sukcharoen N, Keith J, Irvine DS, et al. Prediction of the in-vitro fertilization (IVF) potential of human spermatozoa using sperm function tests: the effect of the delay between testing and IVF. Hum Reprod. 1996;11:1030.
94. Hammadeh ME, Radwan M, Al-Hasani S, et al. Comparison of reactive oxygen species concentration in seminal plasma and semen parameters in partners of pregnant and non-pregnant patients after IVF/ICSI. Reprod Biomed Online. 2006;13:696.
95. McKinney KA, Lewis SE, Thompson W. Reactive oxygen species generation in human sperm: luminol and lucigenin chemiluminescence probes. Arch Androl. 1996;36:119.
96. Aitken RJ, West KM. Analysis of the relationship between reactive oxygen species production and leucocyte infiltration in fractions of human semen separated on Percoll gradients. Int J Androl. 1990;13:433.
97. Saleh RA, Agarwal A, Kandirali E, et al. Leukocytospermia is associated with increased reactive oxygen species production by human spermatozoa. Fertil Steril. 2002;78:1215.
98. Aitken RJ, Buckingham DW, West K, et al. On the use of paramagnetic beads and ferrofluids to assess and eliminate the leukocytic contribution to oxygen radical generation by human sperm suspensions. Am J Reprod Immunol. 1996;35:541.
99. Shekarriz M, DeWire DM, Thomas Jr AJ, et al. A method of human semen centrifugation to minimize the iatrogenic sperm injuries caused by reactive oxygen species. Eur Urol. 1995;28:31.
100. Aitken RJ, Baker MA, O'Bryan M. Shedding light on chemiluminescence: the application of chemiluminescence in diagnostic andrology. J Androl. 2004;25:455.
101. Aitken RJ, Clarkson JS. Cellular basis of defective sperm function and its association with the genesis of reactive oxygen species by human spermatozoa. J Reprod Fertil. 1987;81:459.
102. Kobayashi H, Gil-Guzman E, Mahran AM, et al. Quality control of reactive oxygen species measurement by luminol-dependent chemiluminescence assay. J Androl. 2001;22:568.
103. Kodama H, Yamaguchi R, Fukuda J, et al. Increased oxidative deoxyribonucleic acid damage in the spermatozoa of infertile male patients. Fertil Steril. 1997;68:519.
104. Spano M, Bonde JP, Hjollund HI, et al. Sperm chromatin damage impairs human fertility. The Danish First Pregnancy Planner Study Team. Fertil Steril. 2000;73:43.
105. Evenson DP, Jost LK, Marshall D, et al. Utility of the sperm chromatin structure assay as a diagnostic and prognostic tool in the human fertility clinic. Hum Reprod. 1999;14:1039.
106. Duran EH, Morshedi M, Taylor S, et al. Sperm DNA quality predicts intrauterine insemination outcome: a prospective cohort study. Hum Reprod. 2002;17:3122.
107. Sun JG, Jurisicova A, Casper RF. Detection of deoxyribonucleic acid fragmentation in human sperm: correlation with fertilization in vitro. Biol Reprod. 1997;56:602.
108. Lopes S, Sun JG, Jurisicova A, et al. Sperm deoxyribonucleic acid fragmentation is increased in poor quality semen samples and correlates with failed fertilization in intracytoplasmic sperm injection. Fertil Steril. 1998;69:528.
109. Bungum M, Humaidan P, Axmon A, et al. Sperm DNA integrity assessment in prediction of assisted reproduction technology outcome. Hum Reprod. 2007;22:174.
110. Huang CC, Lin DP, Tsao HM, et al. Sperm DNA fragmentation negatively correlates with velocity and fertilization rates but might not affect pregnancy rates. Fertil Steril. 2005;84:130.
111. Gandini L, Lombardo F, Paoli D, et al. Full-term pregnancies achieved with ICSI despite high levels of sperm chromatin damage. Hum Reprod. 2004;19:1409.
112. Jurisicova A, Varmuza S, Casper RF. Programmed cell death and human embryo fragmentation. Mol Hum Reprod. 1996;2:93.
113. Tesarik J, Mendoza C, Greco E. Paternal effects acting during the first cell cycle of human preimplantation development after ICSI. Hum Reprod. 2002;17:184.

114. Sakkas D, Mariethoz E, St John JC. Abnormal sperm parameters in humans are indicative of an abortive apoptotic mechanism linked to the Fas-mediated pathway. Exp Cell Res. 1999;251:350.
115. Sakkas D, Moffatt O, Manicardi GC, et al. Nature of DNA damage in ejaculated human spermatozoa and the possible involvement of apoptosis. Biol Reprod. 2002;66:1061.
116. Aoki VW, Moskovtsev SI, Willis J, et al. DNA integrity is compromised in protamine-deficient human sperm. J Androl. 2005;26:741.
117. Iguchi N, Yang S, Lamb DJ, et al. An SNP in protamine 1: a possible genetic cause of male infertility? J Med Genet. 2006;43:382.
118. Tremellen K. Oxidative stress and male infertility a clinical perspective. Hum Reprod Update. 2008;14:243.
119. Greco E, Iacobelli M, Rienzi L, et al. Reduction of the incidence of sperm DNA fragmentation by oral antioxidant treatment. J Androl. 2005;26:349.
120. Smith R, Kaune H, Parodi D, et al. Increased sperm DNA damage in patients with varicocele: relationship with seminal oxidative stress. Hum Reprod. 2006;21:986.
121. Werthman P, Wixon R, Kasperson K, et al. Significant decrease in sperm deoxyribonucleic acid fragmentation after varicocelectomy. Fertil Steril. 2008;90:1800.
122. Zini A, Blumenfeld A, Libman J, et al. Beneficial effect of microsurgical varicocelectomy on human sperm DNA integrity. Hum Reprod. 2005;20:1018.
123. Smit M, Romijn JC, Wildhagen MF, et al. Decreased sperm DNA fragmentation after surgical varicocelectomy is associated with increased pregnancy rate. J Urol. 2010;183:270.
124. Potts RJ, Newbury CJ, Smith G, et al. Sperm chromatin damage associated with male smoking. Mutat Res. 1999;423:103.
125. Pasqualotto FF, Sharma RK, Potts JM, et al. Seminal oxidative stress in patients with chronic prostatitis. Urology. 2000;55:881.
126. O'Connell M, McClure N, Lewis SE. Mitochondrial DNA deletions and nuclear DNA fragmentation in testicular and epididymal human sperm. Hum Reprod. 2002;17:1565.
127. Greco E, Scarselli F, Iacobelli M, et al. Efficient treatment of infertility due to sperm DNA damage by ICSI with testicular spermatozoa. Hum Reprod. 2005;20:226.
128. Dohle GR. Male infertility in cancer patients: review of the literature. Int J Urol. 2010;17:327.
129. van der Horst-Schrivers AN, van Ieperen E, Wymenga AN, et al. Sexual function in patients with metastatic midgut carcinoid tumours. Neuroendocrinology. 2009;89:231.
130. The practice Committee of the Society for Assisted Reproductive Technology and the Practice Committee of the American Society for Reproductive Medicine. Preimplantation genetic testing: a Practice Committee opinion. Fertil Steril. 2007;88:1497.
131. Helmerhorst FM, Perquin DA, Donker D, et al. Perinatal outcome of singletons and twins after assisted conception: a systematic review of controlled studies. BMJ. 2004;328:261.
132. Jackson RA, Gibson KA, Wu YW, et al. Perinatal outcomes in singletons following in vitro fertilization: a meta-analysis. Obstet Gynecol. 2004;103:551.
133. The ESHRE Capri Workshop Group. Intracytoplasmic sperm injection (ICSI) in 2006: evidence and evolution. Hum Reprod Update. 2007;13:515.
134. Schlegel PN, Shin D, Goldstein M. Urogenital anomalies in men with congenital absence of the vas deferens. J Urol. 1996;155:1644.
135. Anguiano A, Oates RD, Amos JA, et al. Congenital bilateral absence of the vas deferens. A primarily genital form of cystic fibrosis. JAMA. 1992;267:1794.
136. McCallum T, Milunsky J, Munarriz R, et al. Unilateral renal agenesis associated with congenital bilateral absence of the vas deferens: phenotypic findings and genetic considerations. Hum Reprod. 2001;16:282.
137. O'Flynn O'Brien KL, Varghese AC, Agarwal A. The genetic causes of male factor infertility: a review. Fertil Steril. 2010;93:1.
138. Ferlin A, Raicu F, Gatta V, et al. Male infertility: role of genetic background. Reprod Biomed Online. 2007;14:734.

139. Denschlag D, Tempfer C, Kunze M, et al. Assisted reproductive techniques in patients with Klinefelter syndrome: a critical review. Fertil Steril. 2004;82:775.
140. Nielsen J, Wohlert M. Chromosome abnormalities found among 34,910 newborn children: results from a 13-year incidence study in Arhus, Denmark. Hum Genet. 1991;87:81.
141. Engels H, Eggermann T, Caliebe A, et al. Genetic counseling in Robertsonian translocations der(13;14): frequencies of reproductive outcomes and infertility in 101 pedigrees. Am J Med Genet A. 2008;146A:2611.
142. Kuroda-Kawaguchi T, Skaletsky H, Brown LG, et al. The AZFc region of the Y chromosome features massive palindromes and uniform recurrent deletions in infertile men. Nat Genet. 2001;29:279.
143. Hopps CV, Mielnik A, Goldstein M, et al. Detection of sperm in men with Y chromosome microdeletions of the AZFa, AZFb and AZFc regions. Hum Reprod. 2003;18:1660.
144. Kaufman GE, Malone FD, Harvey-Wilkes KB, et al. Neonatal morbidity and mortality associated with triplet pregnancy. Obstet Gynecol. 1998;91:342.
145. Pinborg A. IVF/ICSI twin pregnancies: risks and prevention. Hum Reprod Update. 2005; 11:575.
146. Klemetti R, Gissler M, Sevon T, et al. Children born after assisted fertilization have an increased rate of major congenital anomalies. Fertil Steril. 2005;84:1300.
147. Reddy UM, Wapner RJ, Rebar RW, et al. Infertility, assisted reproductive technology, and adverse pregnancy outcomes: executive summary of a National Institute of Child Health and Human Development workshop. Obstet Gynecol. 2007;109:967.
148. Hansen M, Kurinczuk JJ, Bower C, et al. The risk of major birth defects after intracytoplasmic sperm injection and in vitro fertilization. N Engl J Med. 2002;346:725.
149. Zhu JL, Basso O, Obel C, et al. Infertility, infertility treatment, and congenital malformations: Danish national birth cohort. BMJ. 2006;333:679.
150. Koivurova S, Hartikainen AL, Gissler M, et al. Neonatal outcome and congenital malformations in children born after in-vitro fertilization. Hum Reprod. 2002;17:1391.
151. Ludwig M, Katalinic A. Malformation rate in fetuses and children conceived after ICSI: results of a prospective cohort study. Reprod Biomed Online. 2002;5:171.
152. Anthony S, Buitendijk SE, Dorrepaal CA, et al. Congenital malformations in 4224 children conceived after IVF. Hum Reprod. 2002;17:2089.
153. Ericson A, Kallen B. Congenital malformations in infants born after IVF: a population-based study. Hum Reprod. 2001;16:504.
154. Sutcliffe AG, D'Souza SW, Cadman J, et al. Minor congenital anomalies, major congenital malformations and development in children conceived from cryopreserved embryos. Hum Reprod. 1995;10:3332.
155. Olson CK, Keppler-Noreuil KM, Romitti PA, et al. In vitro fertilization is associated with an increase in major birth defects. Fertil Steril. 2005;84:1308.
156. Lie RT, Lyngstadaas A, Orstavik KH, et al. Birth defects in children conceived by ICSI compared with children conceived by other IVF-methods; a meta-analysis. Int J Epidemiol. 2005;34:696.
157. Hansen M, Bower C, Milne E, et al. Assisted reproductive technologies and the risk of birth defects a systematic review. Hum Reprod. 2005;20:328.
158. Bonduelle M, Ponjaert I, Steirteghem AV, et al. Developmental outcome at 2 years of age for children born after ICSI compared with children born after IVF. Hum Reprod. 2003;18:342.
159. Zadori J, Kozinszky Z, Orvos H, et al. Dilemma of increased obstetric risk in pregnancies following IVF-ET. J Assist Reprod Genet. 2003;20:216.
160. Simpson JL, Lamb DJ. Genetic effects of intracytoplasmic sperm injection. Semin Reprod Med. 2001;19:239.
161. Alukal JP, Lipshultz LI. Safety of assisted reproduction, assessed by risk of abnormalities in children born after use of in vitro fertilization techniques. Nat Clin Pract Urol. 2008;5(3): 140–50.

Chapter 6
Update on Sperm Banking

Pankaj Talwar

Long life and desire to leave genetic footprints in this world has been a long cherished dream of human mankind. The concept of sperm banking has fulfilled these dreams partly by being able to preserve body cells and genital tissues. The banking of the male gametes involves initial exposure of the spermatozoa to the cryoprotectant and gradually cooling them to subzero temperatures as per desired cooling curve [1]. Semen sample in suitable container is then cryopreserved in liquid nitrogen at $-196°$ centigrade for later use. Cryobioreposited semen sample is then thawed, gradually warmed to the room temperature and diluted with suitable buffered media. Cryoprotectant is washed away and the thawed sample evaluated and used for insemination, intracytoplasmic sperm injection (ICSI) or for research purposes.

The male gamete must maintain its macro and microarchitectural structure along with genomic integrity during the whole procedure and recover its physiological functions completely after the procedure [2].

Factors known to effect outcome of this delicate procedure depends upon the quality of the semen specimen, developmental stage at which sperms are being frozen, type of cryoprotectant being used, and the freezing protocols [3].

Background of Sperm Banking

Freezing and thawing techniques have been used to cryopreserve semen since 1776 [4]. The newer techniques have been developed with scientific advancements in measurement of temperature and the chemistry of cryopreserving solutions. Equally important were innovations made in the nineteenth century involving understandings

P. Talwar, M.D. (✉)
ART Centre, Army Hospital Research and Referral,
Dhaula Kuan, New Delhi 110010, India
e-mail: pankaj_1310@yahoo.co.in

E. Seli and A. Agarwal (eds.), *Fertility Preservation in Males: Emerging Technologies and Clinical Applications*, DOI 10.1007/978-1-4614-5620-9_6,
© Springer Science+Business Media New York 2012

of liquefaction of gases and potential use of such refrigerants to cool and store specimens at extremely low temperatures.

Advancements in the equipments, disposables, and awareness about sterility have also helped the cryobiologist and embryologist to further enhance the freeze thaw cycle results. This approach of cryopreserving human semen has become an acceptable procedure in assisted reproductive technology (ART) laboratories across the world.

Few of the developmental mile stones are worth mentioning:

1776 – Spallanzani – observed the effects of freezing on human sperm and subsequent recovery of motility on warming.

1866 – Montegazza – proposed semen banking for veterinary practice and for soldiers going to battle field [5].

1930 – Shettels and Jahnel S – observed that sperm survives at temperatures as low as −269°C [6, 7].

1949 – Polge, Smith, and Parkes – accidental discovery of cryoprotective properties of glycerol [8].

1963 – Sherman – reported birth of first child born after insemination by sperms which had been frozen using liquid nitrogen [9].

Principles of Cryobiology

Cryopreservation is a process which maintains cellular life for an extended period of time at low subzero temperatures. The aim of any cryopreservation protocol is to minimize cell membrane damage associated with exposure to low temperatures, regulate cell volume during the procedure, and prevent lethal intracellular ice crystal formation. This can be achieved by controlling intracellular and extracellular movement of solutes and water [10]. The outcome of freeze thaw cycle depends on various factors:

1. The cryoprotectant (extender) in which the cells are suspended during cooling and warming.
2. The rate at which the cells are cooled.
3. The temperature at which the sample is plunged into liquid nitrogen.
4. The warming rate.
5. Cryoprotectant removal after thawing.

The major physical and chemical events of cryopreservation that lead to cell membrane damage, include the loss of water from the cell due to osmolarity of the cryopreservation medium, formation of intracellular and extracellular ice crystals, the resultant increase in osmolality of the remaining liquid and reversal of the process during thawing. At least half of the motile sperm sustains cryoinjury when subjected to cryopreservation and thawing [11].

The typical effects observed in frozen-thawed spermatozoa derive from the unique nature of the cell. Human spermatozoa are relatively simple cells with a large surface area/volume ratio and have high permeability to water. This ensures rapid osmotic equilibrium in the presence of cryoprotectants. The genetic material is highly condensed and is less prone to injury to cryoprotectants unlike embryos.

Biological Behavior of the Phospholipids Membrane of the Sperm

Cell membranes of all mammalian species are composed of a phospholipid bilayer and associated membrane proteins (receptors, enzymes, etc.). Phospholipid bilayers are composed of a polar head group with hydrocarbon tails. Specific phospholipids polar heads face outward and others face inward, with hydrocarbon tails directed inward from each polar head group. Some nonbilayer lipids are associated with specific integral proteins (receptors, enzymes, etc.). This association influences the protein's functional properties [12].

Cold shock etiology involves damage to the cellular membranes and alteration in metabolic function, probably caused by changes in the arrangement of membrane constituents. When a cell membrane is cooled, a reordering of the membrane components occur which leads to increased membrane viscosity and decreased fluidity. A decrease in temperature causes a thermotropic phase transition in the membrane phospholipids from a liquid-crystalline to a gel phase, resulting in more rigid membrane structure. The occurrence of cold shock can be prevented by controlling the rate of cooling and by adding protective compounds to semen diluents [13–15].

When nonbilayer lipids associated with the membrane proteins reach their transition temperature, a phase separation occurs within the plane of the membrane, which contains the gel-forming lipids. Lipids-associated proteins may be excluded laterally into areas of the membrane, which are still in a liquid state. The movement of the proteins within the membrane leads them to associate with different membrane lipids, possibly changing the activity of the protein. Rewarming of the membrane also could result in altered lipid/protein patterns.

Changes in Volume of Sperm When Exposed to Cryoprotectants

Spermatozoa undergo several volume changes during cryopreservation and thawing. The changes in cell volume throughout freezing and thawing are reliant upon water transport across the cell's plasma membrane. The first volume change occurs when the cryoprotective agent (CPA) is added to the sperm suspension.

The sperm undergoes rapid shrinkage as intracellular water leaves the cell in response to the hyperosmotic extracellular solution. Gradually, the sperms returns to its original volume as the CPA permeate the cell. The presence of CPA in the intracellular and extracellular water lowers the freezing point of the cell and the solutes, allowing them to remain in super cooled state.

Ice formation is initiated in the extracellular medium. As more water leaves the cryopreservation medium to contribute to crystal formation, the concentration of particles in the remaining liquid increases, resulting in a continuous drop in the freezing point of the remaining liquid and an increase in extracellular osmotic pressure. With rising hyperosmolarity of the extracellular fluid, more intracellular water is drawn out of the spermatozoa leaving the cell dehydrated and reducing the risk of lethal intracellular ice formation [16].

Indications of Semen Cryopreservation

Sperm banking is the process of semen cryopreservation using well-documented protocols for future use (Table 6.1). The semen is either preserved by an individual for his own future use, which is termed as autologous sperm banking or client depositor semen cryofreezing, or by fertile donors, after screening, for the purpose of third party reproduction [17, 18]. Adequate care is taken to do phenotypic/blood group matching in these cases. Matching physical characteristics and race of the partner, hair color, texture, and eye color are mandatory (Table 6.2).

Common Indications for Autologous Semen Banking

- Malignancy patients prior to surgery, chemotherapy, or radiation therapy.
- During the performance of cold surgical sperm retrieval (SSR) techniques like testicular or epididymal aspiration of sperms in cases with azoospermia or having the inability to achieve ejaculation.
- In case of anticipated absence of an individual on the day of insemination, such as soldiers, frequent travelers and individuals with professional and other commitments.
- Anticipated performance anxiety on the day of IUI/IVF.

Table 6.1 Current approach to male fertility preservation

Clinical presentation	Modality of sperm harvesting	Modality of banking
Infertile normal reproductive males requiring semen banking	Masturbation/electro-ejaculation	Semen freezing
Infertile azoospermic males		
Obstructive	Sample extracted from the epididymis – percutaneous epididymal sperm aspiration (PESA)	Freezing of the epididymal aspirate either unprepared or after gentle wash and swim up
Nonobstructive	Testicular sperm extraction/ aspiration (TESE/TESA)	Freezing of the testicular tissue is done after gentle teasing in a sterile dish using fine needles
		Tissue is frozen either unprepared or after density gradient wash
In malignancy patients before chemotherapy or radiotherapy		
Prepubertal boys	Masturbation	Semen banking
Pubertal	Testicular tissue (multiple samples)	Freezing of the testicular tissue

Table 6.2 Indications of semen cryopreservation

Donor semen banking	Autologous semen banking
Azoospermia (hypergonadotropic – hypogonadism)	Adult patients prior to surgery, chemotherapy, or radiation therapy
Very low sperm count and couple willing for donor insemination	Soldiers/frequent travelers before they go away for overseas assignments, or due to anticipated absence on day of insemination due to commitments of work
Male partner carries a genetic defect (Huntington's Chorea, etc.)	During performance of cold surgical sperm retrieval (SSR) techniques like testicular or epididymal aspiration of sperms in cases with azoospermia
Single women donor insemination	Anticipated performance anxiety on the day of assisted reproduction technology procedure
Recurrent IVF failures/abortions	Patients before undergoing vasectomy may preserve semen as insurance to further fertility
	Spinal cord injured patient: electroejaculation specimen can be cryopreserved and successfully used in assisted reproduction for spinal cord injured patient
	Oligozoospermic/asthenozoospermic sample: Anticipation of enriching the number of motile sperm by pooling several cryopreserved samples

- Semen retrieval by electro-ejaculation in men suffering from spinal cord injury.
- Occupational hazard (risk of radiation).
- Patients before undergoing vasectomy may preserve semen as insurance to further fertility.
- Pooling of semen sample when the native semen samples have low count.

Indications for Donor Semen Banking

- Male factor infertility
- Azoospermia (hypergonadotropic hypogonadism)
- Morphological abnormalities of the sperms.
- Very low sperm count and couple willing for IUI-AID.
- Partner is either impotent or has retrograde ejaculations (if sperm retrieval has failed).

  Male partner carries a genetic defect (Huntington's Chorea, etc.).
  Rh incompatibility with isoimmunization.
  Single women donor insemination.
  Recurrent IVF failures/abortions.
  Recent indication desirous of designer baby.

  When exposed to known toxins, such as lead and agents with mutagenic potential.

Outline of Cryoprotectants

Biochemical and Physical Aspects of Sperm Cryopreservation

Sperms subjected to cryopreservation and thawing have reduced viability, motility and fecundity. Some of the morphological damages to the membranes and acrosome associated with cryopreservation and thawing are increased plasma membrane permeability, loss of acrosomal integrity and loss of superoxide dismutase enzyme from the plasma membranes [19, 20].

All the cryopreservation protocol are based on the theory that cell membrane damage can be minimized through the addition of suitable CPAs, buffer agents, controlling the osmolality and pH of cryopreservation medium and controlling the freezing rate during the procedure [21].

A variety of extenders (cryoprotective media) exist for the cooling and cryopreservation of semen. The purpose of the extender is manifold. The media contains nutrient, a buffer, a cryoprotectant agent, and an antibiotic. A typical nutrient added is a sugar, such as glucose or sucrose, which serves to provide energy source for the sperm. Buffers are added to balance pH and osmolarity of the solution. An ideal biological buffer should have a pH value between 6 and 8 [22]. The role of the buffer in cryopreservation is to pick up hydrogen ions in the surrounding media, thereby assisting in dehydration of the cell and maintaining a neutral pH.

Basics of Cryoprotectants as Applied to Semen Banking

Cryoprotectants have been divided into two classes, with first being permeating cryoprotectants, such as dimethylsulfoxide (DMSO), propylene glycol, and glycerol, and the second being non-permeating cryoprotectants, such as sucrose, raffinose and glycine.

Glycerol has remained the cryoprotectant of choice for the preservation of spermatozoa for most species [23]. Glycerol is superior to DMSO or ethylene glycol as a cryoprotectant [24]. The protective effects of glycerol are mediated by its colligative properties, depression of freezing point, alteration of cell membrane properties by inducing changes in lipid packing structure, and the consequent lowering of electrolyte concentrations in the unfrozen fraction at any given temperature, which will help to counter the harmful "solution effects" imposed during the freezing process [25]. In order to improve cryosurvival rates, more complex diluents containing other mainly nonpermeable CPAs, such as glycine, zwitterions, citrate, and egg yolk were developed [26]. Among the earliest and best known extenders for human semen is glycerol egg yolk citrate (GEYC). Lipoproteins and phospholipid fractions of egg yolk were originally added to cryopreservation media in effort to protect the sperm from the deleterious effects of cold shock by protecting the integrity of the plasma membrane. Studies have demonstrated the binding of various lipid constituents of egg yolk to the spermatozoa membrane [27]. Commonly used human sperm preserving medium (HPSM), is a modified Tyrode's medium containing glycerol (5–7.5% v/v final volume), sucrose, glucose, and glycine as CPAs, human serum albumin as stabilizing agent and (N-(2-hydroxyethyl)-piperazine-N'-(2-ethanesulfonic acid)) (HEPES) as the buffering agent. Other commonly used cryoprotective buffer is a zwitterion buffer system containing N-tris-(hydroxymethyl)-methyl-2-aminoethane sulfonic acid (TES) and tris-(hydroxymethyl)-aminomethane (TRIS). This TES-TRIS combination (usually abbreviated to "TEST") is most often used in conjunction with egg yolk and citrate with glycerol as the permeating cryoprotectant. In the initial report, it proved superior to glycerol alone as a cryoprotectant. This remarkable result was due largely to the very rapid freezing protocol employed, and was impossible to duplicate using standard cooling methods [28].

The addition of the cryoprotectant glycerol has been shown to increase the motile sperm cryosurvival to an average of 50%, well above the <20% cryosurvival reported without addition of glycerol. The cryoprotectant DMSO was shown to be unsuitable for sperm cryopreservation as it had lower post-thaw percent motility when compared to glycerol. Numerous studies have reported irreversible reduction in sperm motility after exposure to glycerol and other intracellular CPAs [29].

Essentials of Freeze Thaw Cycle

Decision of Semen Packaging Protocol Before the Procedure and Its Importance to the Reproductive Biologist

Semen parameters should be thoroughly assessed using WHO criterion. We should decide the outcome of the procedure which will further guide the freezing protocol. If the sample is satisfactory with good count and motility, we may freeze raw sample with the aim to carry out IUI at latter date. On the other hand, an oligospermic sample with round cells and debris should be prepared and packaged for ICSI. Finally, the methodology depends upon the experience of the reproductive biologist and the laboratory protocols.

Specimen Glycerolization

Glycerol is added directly or indirectly, as a component of a cryopreservation medium, to neat semen/prepared semen (swim up) in drop by drop fashion slowly over a period of 2–3 min, with continuous mixing of both. This step is essential to reduce toxicity of the cryoprotectant as it can cause sudden osmotic shock to the sperms. The glycerol is metabolized during the procedure with the formation of neutral lipid. It is suggested that the metabolized glycerol may contribute to the plasma membrane of the sperm increasing its stability which may lead to improved post-thaw motility [30].

Packaging of Semen After Addition of Cryopreservative

Glycerated or extended semen can be cryopreserved in various containers. Factors, which influence the decisions, include the volume of sample to be cryopreserved, ease of container labeling, handling, storage, and recovery as well as biocompatibility of the packaging material. The packaging of semen into traditional 0.25 or 0.5 mL straws requires the use of vacuum pump and filling nozzle to aspirate the semen-cryoprotectant mixture. Straws made from an ionomeric resin (CBS High Security Straws, commonly referred to as "CBS straws": CryoBioSystem, Paris, France) have shown encouraging results. Straws are available in a variety of colors suitable for the easy identification of samples, and many hundreds can be stored in plastic goblets in canisters within liquid nitrogen vessels. Disadvantages of straws include: (1) Maximum capacity of approximately 0.5 mL only. (2) Overfilled straws are prone to cracking and expelling the powder sealing plugs into the liquid nitrogen. (3) Labeling difficulties. (4) Filling difficulties. (5) A high surface/volume ratio which makes the sample very susceptible to warming shock damage resulting from exposure to ambient temperatures during handling

Cryovials made from polypropylene with either polypropylene or polyethylene screw caps have the advantages of ease of labeling and handling. These are easy to fill and store nearly 1.5 mL of the semen plus cryoprotectant mixture. Storage on aluminum canes is not good as after sometime aluminum canes lose their memory and cryovials may jump off the holder. These occupy lots of cryocan space and are not good for busy semen bank. Some amount of leakage does occur which can be prevented by the use of NUNC™ CryoFlex™ tubes. Due to more capacity, have the potential of exploding upon thawing because liquid nitrogen trapped in the vials expand to many times its volume when it converts to gaseous nitrogen. The low surface: volume ratio and thick wall of the cryovial increase the time required for samples to reach critical temperatures and thus increase the risk of damage from brief exposure to ambient temperatures. Placing cryovials in drawers in racking system is more efficient, but prone to induce fluctuation in storage temperature during retrieval. Screw-top vials do not maintain their seals, and leakage of liquid in nitrogen into a container is common, with consequent risk of rupture during thawing. *Glass vials* transmit heat more uniformly and were preferred until safety considerations for embryologist and andrologist made it preferable to use plastic straws or vials. Glass vials are very fragile thus there use is discouraged these days. *Syringes* are difficult to be loaded and sealed safely and aseptically. It has the advantage for direct insemination, but this is outweighed by the excessive amount of space needed to store multiple ejaculates.

Cooling and Warming Rates

The outcome of sperm cryosurvival is related to the rate at which the cells are cooled and warmed. A high rate of cooling does not allow time for the sufficient amount of intracellular water to diffuse out of the cells. These can form ice crystals during further cooling and cause intracellular damage. During warming, these small crystals aggregate, forming larger crystals, which is called recrystallization. Formation of these large crystals is believed to cause cell damage or death by physical disruption of the plasma membrane and/or cell organelle membranes. A relatively slow cooling rate allows excess intracellular water to diffuse out of the cells, resulting in extreme cell shrinkage, cell dehydration, and a high intracellular solute concentration. The risk of intracellular ice formation is not completely eliminated by slow cooling. While slow cooling reduces the risk of lethal intracellular ice formation, extreme changes in cell volume and high intracellular solute concentrations may damage the cell membrane [31].

Freezing

Sperm cryopreservation is accomplished using liquid nitrogen vapors for noncontrolled rate or a programmable freezer for controlled rate cooling. Comparisons of the controlled rate and noncontrolled rate freezing have revealed no differences in sperm cryosurvival or post-thaw motility. Noncontrolled rate freezing in liquid nitrogen vapor is simpler and less expensive. Regardless of the cooling process, the ultimate quality control appraisal of sperm cryopreservation is the cryosurvival of the spermatozoa determined during thawing [32].

Storage

Once specimens attain temperature of −80°C to −120°C, they are immediately plunged into liquid nitrogen. After plunging, the vials are quickly transferred to a precooled, labeled aluminum cane/goblet for storage in liquid phase of liquid nitrogen tank. When freezing semen by noncontrolled rate protocol, vials may be loaded on to the aluminum cane prior to cryopreservation to eliminate the need for transfer after cryopreservation. Every effort should be made to limit the time; specimens spend out of the liquid phase of liquid nitrogen.

Straws are quickly transferred to precooled, labeled plastic goblets, snapped onto a labeled aluminum cane. Straws should be oriented in the goblet so that identifying information can be read without completely withdrawing the straw from the goblet. Aluminum canes are placed in predetermined locations within the cryostorage vessel. All storage containers should be stored in a secured room in locked/chained containers. Liquid nitrogen dewars and storage tanks are available in a variety of sizes. Dewars require manual filling while most storage tanks have an automatic filling feature. Liquid nitrogen levels in storage units should be monitored regularly at all times. It is important to appreciate the length of time cryopreserved sperm may be stored for. At −196°C, storage of sperms, even for a lengthy period of time, does not affect the survival rates. Liquid nitrogen holds specimens at a temperature (−196°C) at which there is virtually no movement of atoms or molecules. At temperatures above −130°C, atoms, and molecules are able to move. Temperatures of −90°C and above allow ice crystal growth and even short periods of exposure to such temperatures can cause lethal damage to cells. As long as the cells are maintained at −196°C, the only known potential for cell damage is degradation of deoxyribonucleic acid (DNA) caused by background radiation

Based on normal background radiation of 0.1 rad/year, it has been predicted that the male gametes should maintain its genetic integrity for over 200 years when stored at −196°C.

Thawing

Practical approach to semen thawing would be to wash the vials/straws in running water. They should be cleaned externally till the sweating disappears over a period of few minutes. Once the sample is thawed, mix the sample well with a pipette before sampling. Perform the sperm count and assess the motility as per WHO guidelines. Specimen should be processed immediately after post-thaw analysis.

Effects of Cryofreeze/Thaw Cycle

The success of cryopreservation is measured by the number of motile spermatozoa recovered post thaw. There is nearly 30–40% loss in motility in the thawed sample. Sperm membranes are destabilized by the sudden transitions of the temperature (thermal stress), the volume alterations with water and cryoprotectant shifts and subsequent exposure to changing osmolarity (osmotic stress). At a molecular level, changes in membrane organization, such as modifications of specific lipid-protein interaction, phospholipid asymmetry, and lipid composition are implicated in the loss of permeability and rigidifying effect on the membrane fluidity. Cryopreservation has deleterious effects on spermatozoa, especially on plasmalemma, acrosomes and chromatin integrity [33]. Studies have compared the recovery of postthaw motility obtained with the various cryoprotective extenders and some of their derivatives. Results are conflicting with no obviously superior candidate emerging, but this is probably a reflection of the various cooling and thawing rates employed by the different research groups, making comparison difficult. It is recommended that additional tests of functional/fertilization ability of the sperms, such as cervical mucus penetration or zona-free hamster oocyte fusion testing, are applied to assess the potential fertility of cryopreserved spermatozoa [34].

Legislation Pertaining to the Semen Banking

Indian Council of Medical Research has issued comprehensive guidelines for assisted reproductive technology centers. These are guidelines which may be applied to any functioning semen bank. An ART clinic or a law firm or any other suitable independent organization may set up a semen bank. All donors should produce their semen samples within the collection area of the center so that the sample cannot be substituted by the semen sample of others. It is essential that there is suitable privacy, time, and environment for patients to do this. Donor records and coding of the specimens stored must be kept securely. The centers should audit their cryobanks annually. As per the ICMR guidelines, the semen bank shall not supply semen of one donor for more than ten successful pregnancies. It will be the responsibility of the ART clinic or the patient, to inform the bank about a successful pregnancy. The bank shall keep a record of all semen received, stored, and supplied, and details of the use of the semen of each donor. This record will be liable to be reviewed by the accreditation authority. A semen bank may store a semen preparation for exclusive use on the donor's wife or on any other woman designated by the donor. An appropriate charge may be levied by the bank for the storage. In the case of nonpayment of the charges when the donor is alive, the bank would have the right to destroy the semen sample or give it to a bonafide organization to be used only for research purposes. In the case of the death of the donor, the semen would become the property of the legal heir or the nominee of the donor at the time the donor gives the sample for storage to the bank. In the UK, the semen is not normally stored for longer than 10 years or beyond the age of 55 years for donors; generally, it is not stored in France for more than 5 years. Donors may express a wish to further limit the period of storage or the number of pregnancies that can be obtained from one donor. This is restricted to ten children by the same donor. Confidentiality remains a primary consideration in most countries.

Donor Screening Prior to Semen Banking

Fresh donor insemination is not recommended for the fear of transmission of common infective diseases. Donors should be tested for HIV 1 and 2, HTLV I and II antibodies, hepatitis B surface antigen, hepatitis B core antibody, hepatitis C, RPR, TP-PA, cytomegalovirus antibodies, chlamydia and gonorrhea. Some agents and diseases that can be transmitted by the seminal fluid include HIV, hepatitis B, hepatitis C, and syphilis.

Donors are screened for the infections at the time of presentation. As a donor may be in window period of an infection, it is necessary to repeat the examination for hepatitis B and HIV after an appropriate quarantine period of 180 days. If the history or physical examination indicates infection, the donor should be rejected and advised to seek appropriate medical advice. Donor should be thoroughly screened for common genetic and communicable diseases and those specific to their geographic location before their enrolment in the program.

Cross-Infection in the Semen Banks

There is a potential danger of cross-infection within the bank thus the samples must be handled and stored with paramount care. Cryopreserved semen may be spilled in the cryocan, and the infectious organism (hepatitis B virus) may survive in the liquid nitrogen with the possibility of cross-infection of other stored samples. It is recommended that samples be stored in isolation cryocans till the quarantine period of HIV, HbsAg, and HCV. Use of CBS ionomeric straws which have been hermetically sealed offers protection against cross-infection. Men with malignancy often need to bank their semen at short notice as to preclude complete prefreeze infective screening. The semen of these men could be reposited in a quarantine tank until the requisite screening had been completed. Client depositor/autologous who has the comfort of time, e.g., men considering a vasectomy, a cryobank must insist upon screening for infective pathogens as a safety precaution for the security of the semen of other men stored in the same canister. When cryopreserving semen from patients who are known to carry an infective infection, e.g. HIV and HBsAG, the samples must be stored in separate "contaminated" tanks for each recognized pathogenic organisms. In the USA and in the UK, guidelines have been published. The HEFA is moving toward a position whereby laboratories are compelled to screen donors in this way [35].

Security of the Semen Bank

The straws or vials must be clearly labeled. Inventory control is of utmost importance. Every precaution must be made to ensure that each straw or vial can be linked to the sperm source, date of cryopreservation and specimen number, canister/cane or rack/cryocan number. The cryocans have a locking facility which must be utilized and limited staff allowed access to the keys. A lot of softwares is available in the industry for easy identification of the specimen. Some examples of vial or straw identification mechanisms currently employed by sperm banks include: computerized or manual bar coding system, color coding with cryomarkers, vial caps or straw plugs, or use of adhesive labels.

Secure cryopreservation of semen requires regular maintenance of the equipment and refilling of liquid nitrogen in the cryocans. Liquid nitrogen evaporates very quickly or the cans can leak thus causing loss of precious samples. The loss of stored semen from cancer or other patients like those having spinal cord injuries and in whom semen has been retrieved with electroejaculation is not only difficult to quantify but is also of immense emotional value to the owner/couple. All the details and records must be stored confidentially and country specific guidelines adhered to when carrying out cryofreezing at the center. The following measures are considered customary by various accreditation authorities:

1. The levels of LN2 in cryocans that are filled manually should be monitored on a regular basis, using a cryoscale.

2. Large cryobiorepository cans should be attached to an automated cryomanifold with "auto-fill" controller.
3. Low-level temperature sensors should be installed in all cryogenic storage tanks and also, an alarm should be connected to the tanks to alert biologist about unwarranted problems.

The Future of Semen Cryopreservation

Cryopreservation of human gametes expose them to numerous stresses, including mechanical, thermal, and chemico-physiological, which can lead to compromised function of the gametes. We have come a long way since accidental discovery of glycerol as cryoprotectant for human sperms. The enigma of finding an ideal cryoprotectant would be the holy grail of the modern cryobiology. The applications of semen cryopreservation which had humble origin in veterinary practice now entails cryopreservation of stem cell-derived male sperms, though in experimental stages using vitrification. The introduction of clinical ICSI in 1992 at Brussels opened a new era in the field of Assisted Reproductive Techniques, allowing couples with severe male factor infertility to hope for a child of their own genetic origin. Nearly 2 years later, it was established that pregnancy was possible by carrying out ICSI on sperms extracted by SSR techniques. Nagy et al. showed that comparable result in terms of pregnancy rates can be obtained by performing ICSI with ejaculated, epididymal, or testicular spermatozoa, although fertilization rates were significantly higher with the ejaculated sperm. Epididymal spermatozoa can be retrieved either by microsurgery or by percutaneous needle puncture. These can be subsequently cryofreezed [36, 37]. The frequent indication for epididymal aspiration is obstructive azoospermia, and thus it is not uncommon for relatively large quantities of sperm to be obtained and subsequently used for IVF, or even intrauterine insemination (IUI). Surplus sperm may be frozen for future use. If large numbers of epididymal spermatozoa are obtained, then density gradient centrifugation is an effective method for preparing those spermatozoa for subsequent use. Testicular specimens are contaminated invariably with large amounts of red blood cells and testicular tissues; additional steps are needed to isolate a clean preparation of spermatozoa. In order to free the seminiferous tubulebound spermatozoa, it is necessary to use either enzymatic digestants (collagenase) or mechanical methods. For the latter, testicular tissues in supportive culture medium is macerated using glass coverslips until a fine slurry of dissociated tissues is produced, and the resulting suspension can then be processed for therapeutic use.

Excess testicular spermatozoa obtained in this manner can be frozen for future use in order to avoid further surgeries [38, 39]. Testicular spermatozoa can also be obtained from a needle biopsy, although only a small amount of tissue is usually retrieved and the resulting sperm yield is proportionately low. Another method is cryopreservation of single human spermatozoa in empty human Zona shell, which is established by Jacques Cohen. A hollow sphere remains when cellular material is removed from the zona. Because it can be seen and handled microscopically both before and after cyropreservation, it is an ideal capsule for freezing individual and small groups of sperm cells [40]. The conventional semen freezing techniques have proven inefficient in recovering sperms, from men with severe oligospermia, for IVF. Hence, the concept of cryoprotectant free vitrification of semen sample was mooted. An adequate number of sperms was recovered by ultrafast freezing the prepared sample and then warming them to 37°C using copper loops as carrier device [41, 42]. Semen banking is going to encompass testicular tissue banking in greater number of patients in the years to come.

Emerging Role of Semen Banking in Onco-ART

Malignancy is one of the common diseases with approximately 50% of men facing this diagnosis during the course of their lifetime. Till date, the focus of clinicians has been timely diagnosis and appropriate treatment of the patients. With increasing favorable outcome the focus is shifting to fertility preservation [43, 44]. This, in turn, has provided many patients with the opportunity to live complete lives, allowing them to reflect on life subsequent to treatment fertility issues, such as posttreatment marriage and parenthood are considered as important by many parents and young patients. At the time of cancer diagnosis, patients and clinicians alike are often weighed down by the large number of urgent tests and procedures that must be carried out in a timely manner. In the present scenario, proactivity is desired from the oncology and the Assisted Reproductive Team alike. There is no specific consideration for semen banking in such cases except the time factor and stress they are facing. Sample may be collected as in normal healthy males. While collecting semen for banking from young boys/teenagers, the parents, child, and clinicians face sociopsychological inconvenience due to the nature of the subject and such situations may require psychological counseling [45]. Another problem we face is in young males who are not permitted to masturbate as per the religious guidelines. Such males are advised testicular tissue extraction and banking. Similar protocol is being followed for young cancer patients who do no masturbate. It is recommended to bank multiple vials/straws of the samples as the recovery may not be very good especially in postchemotherapy and or radiotherapy patients. ICSI is the recommended treatment for such patients. Majority of the cancer patients do have sufficient time when they visit us. We should start the procedure on the day of referral itself if possible. Minimum six straws should be frozen from 2 to 3 samples collected 24 h apart. The patient and the parents should be counseled about the modality of treatment at latter date. The chances of offering ICSI should be disused at length. Semen banking is commonly carried out in patients with testicular cancers, Hodgkin's disease and leukemia [46].

Conclusions

Sperm banking has become a widely accepted compendium of infertility treatment and in recent times mooted as having new role as fertility insurance. Once looked upon with improbability, this practice has established time and time again to be a successful technique of keeping the anticipation of a family alive for countless families. The motives for storage are as diverse as humans themselves. So far, no limit has been established for how long human semen can be frozen when maintained and stored in appropriate liquid nitrogen storage. Scientific literature shows conclusively that sperm motility, viability, and morphology are not affected by proper long-term cryopreservation. Cryo thaw semen pregnancies have been reported after two to three decades of semen banking [47, 48]. Appropriate screening should be carried out before semen banking available guidelines. Currently, acceptable guidelines include those by the British Andrology Society [49] and the Practice Committee of the American Society for Reproductive Medicine [50].

See Chap. 12

References

1. Parkes AS. Preservation of human spermatozoa at low temperatures. Br Med J. 1945;2:212–3.
2. Sherman JK, Buge RG. Observations on preservation of human spermatozoa at low temperatures. Proc Soc Exp Biol Med. 1953;82:686–8.
3. Hammerstedt RH, Graham JK, Nolan JP. Cryopreservation of mammalian sperm; what we ask them to survive. J Androl. 1990;11:73–88.
4. Spallanzani L, Opuscoli di Fisca, Amimale, e Vegetgabile, Opuscolo II. Osservazioni, e Speriencze inotrno ai Vermicelli Spermatici dell'Uomo e degli Amimali. Modena; 1776.
5. Montegazza J. Fisologia sullo sperma umano. Rendic reale Instit Lomb. 1866;3:183–5.
6. Shettels LB. The respiration of human spermatozoa and their response to various gases and low temperatures. Am J Physiol. 1940;128:404–15.
7. Hoagland H, Pincus G. Revival of mammalian sperm after immersion in liquid nitrogen. J Gen Physiol. 1942;25:337–44.
8. Polge C, Smith AU, Parkes AS. Revival of spermatozoa after vitrification and dehydration at low temperatures. Nature. 1949;164:666–70.
9. Sherman JK. Research on frozen human semen. Past, present, and future. Fertil Steril. 1964;15:485–99.
10. Meryman HT. Mechanics of freezing in living cell and tissues. Science. 1956;124:515–21.
11. Quinn PH. Principles of membrane stability and phase behaviour under extreme conditions. J Bioenerg Biomemb. 1989;21:3–19.
12. Carruthers A, Melchior DL. Role of bilayer lipids in governing membrane transport processes. In: Aloia RC, Curtin CC, Gordon LM, editors. Lipid domains and the relationship to membrane function. New York: Liss; 1988. p. 201–25.
13. Smith AU, Ploge C, Smiles J. Microscopic observation of living cell during freezing and thawing. J Roy Microsc Soc. 1951;71:186–95.
14. Drobnis EZ, Crowe LM, Berger T, Anchordoguy TJ, Overstreet JW, Crowe JH. Cold shock damage is due to lipid phase transitions in cell membranes: a demonstration using sperm as a model. J Exp Zool. 1993;265:432–7.

15. Holt WV, North RD. Effects of temperature and restoration of osmotic equilibrium during thawing on the induction of plasma membrane damage in cryopreserved ram spermatozoa. Biol Reprod. 1992;46:1086–94.
16. Farrant J. Mechanism of cell damage during freezing and thawing and its prevention. Nature. 1965;205:1284–7.
17. Shapiro SS. Strategies to improve efficiency of therapeutic donor insemination. In: Diamond MP, DeCherney AH, editors. Male infertility, Infertility and reproductive medicine clinics in North America. Philadelphia: WB Saunders; 1992. p. 469–85.
18. Bordson BL, Ricci E, Dicky RP, Dunway H, Taylor SN, Curole DW. Comparison of fecundability with fresh and frozen semen in therapeutic donor insemination. Fertil Steril. 1986;46:466–9.
19. Barthelemy D, Royere D, Hammahah S, Lebos C, Tharanne MJ, Lansac J. Ultrastructural changes in membrane and acrosome of human sperm during cryopreservation. Arch Androl. 1990;25:29–40.
20. Lasso JL, Noiles EE, Alvaraz JG, Storey BT. Mechanism of superoxide dismutase loss from human sperm cell during cryopreservation. J Androl. 1994;15:255–65.
21. Van den Berg L, Soliman FS. Composition and pH changes during freezing of solutions containing calcium and magnesium phosphate. Cryobiology. 1969;6:10–4.
22. Mortimer D. Semen cryopreservation. In: Mortimer D, editor. Practical laboratory andrology. Oxford: Oxford University Press; 1994.
23. Hammit DG, Walker DL, Willamson RA. Concentration of glycerol required for optimal survival and in vitro fertilization capacity of frozen sperm is dependent on cryopreservation medium. Fertil Steril. 1988;49:680–7.
24. Zimmerman SJ, Maude MB, Modldawer M. Freezing and storage of human semen in 50 healthy medical students. A comparative study of glycerol and dimethylsulfoxide as a preservative. Fertil Steril. 1964;15:505–10.
25. Mahadevan M, Trounson A. Effects of cryopreservation media and dilution methods on the preservation of human spermatozoa. Andrologia. 1983;15:355–66.
26. Prins GS, Weidel L. A comparative study of buffer systems as cryoprotectants for human spermatozoa. Fertil Steril. 1986;46:147–9.
27. Ramirez JP, Carreras A, Mendoza C. Sperm plasma membrane integrity in fertile and infertile men. Andrologia. 1992;24:141–4.
28. Jeyendran RS, Van der Ven HH, Kennedy W, PerezPelaez M, Zanelveld LJD. Comparison of glycerol and a zwitterions buffer system as cryoprotective media for human spermatozoa: effect on motility, penetration of zona-free hamster oocytes, and acrosin/proacrosin. J Androl. 1984;5:7.
29. Richardson DW, Sadlier RM. The toxicity of various non-electrolytes to human spermatozoa and their protective effects during freezing. J Reprod Fertil. 1967;14:439–44.
30. Mazur P. Freezing of living cells: mechanisms and implications. Am J Physiol. 1974;247:C125–42.
31. Tyler JPP, Kime L, Cooke S, Driscoll GL. Temperature change in cryo-containers during short exposure to ambient temperatures. Hum Reprod. 1996;11:1510–2.
32. Graham EF, Crabo BG. Some methods of freezing and evaluating human spermatozoa. Proc Natl Acad Sci U S A. 1978;4:274–304.
33. Bunge RG, Keetel WC, Sherman JL. Fertilization capacity of frozen human spermatozoa. Nature. 1953;172:767–8.
34. Mclaughlin EA, Ford WCL, Hull MGR. The contribution of the toxicity of a glycerol-egg yolk-citrate cryopreservative to the decline in human sperm motility during cryopreservation. J Repord Fertil. 1992;95:749–54.
35. Human Fertilization and Embryology Act. Schedule 3 para 2(2)(b). London: HMSO; 1990.
36. Oates RD, Lobel SM, Harris D, et al. Efficacy of intracytoplasmic sperm injection using intentionally cryopreserved epididymal sperm. Hum Reprod. 1996;11:133–8.
37. Elnaser TA, Rashwan H. Testicular sperm extraction and cryopreservation in patients with non-obstructive azoospermia prior to ovarian stimulation for ICSI. Middle East Fertil Soc J. 2004;9(2):128–35.

38. Oates RD, Mulhall J, Burges C, Cunniingham D, Carson R. Fertilization and pregnancy using intentionally cryopreserved testicular tissue as the sperm source for intracytoplasmic sperm injection in 10 men with non-obstructive azoospermia. Hum Reprod. 1997;12:734–9.
39. Gil-Salom M, Romero J, Minguez Y, et al. Pregnancies after intracytoplasmic sperm injection with cryopreserved testicular tissue. Hum Reprod. 1996;11:1309–13.
40. Cohen J, Garrisi GJ, Congedo-Ferrara TA, Kieck KA, Schimmel TW, Scott RT. Cryopreservation of single human spermatozoa. Hum Reprod. 1997;12:994–1001.
41. Isachenko V, Isachenko E, Katkov II, Montag M, Dessole S, Nawroth F, et al. Cryoprotectant-free cryopreservation of human spermatozoa by vitrification and freezing in vapor: effect on motility, DNA integrity, and fertilization ability. Biol Reprod. 2004;71(4):1167–73.
42. Isachenko E, Isachenko V, Katkov II, et al. DNA integrity and motility of human spermatozoa after standard slow freezing versus cryoprotectant-free vitrification. Hum Reprod. 2004;9(4):932–9.
43. Tomlinson MJ, Pacey AA. Practical aspects of sperm banking for cancer patients. Hum Fertil. 2003;6(3):100–5.
44. Wallace WH, Anderson RA, Irvine DS. Fertility preservation for young patients with cancer: who is at risk and what can be offered? Lancet Oncol. 2005;6(4):209–18.
45. Saito K, Suzuki K, Iwasaki A, Yumura Y, Kubota Y. Sperm cryopreservation before cancer chemotherapy helps in the emotional battle against cancer. Cancer. 2005;104(3):521–4.
46. Schmidt KL, Carlsen E, Andersen AN. Fertility treatment in male cancer survivors. Int J Androl. 2007;30(4):413–9.
47. Horne G, Atkinson AD, Pease EH, et al. Live birth with sperm cryopreserved for 21 years prior to cancer treatment: case report. Hum Reprod. 2004;19(6):1448–9.
48. Feldschuh J, Brassel J, Durso N, Levine A. Textbook of assisted reproductive technologies. Successful sperm storage for 28 years. Fertil Steril. 2005;84(4):1017.
49. British Andrology Society. British Andrology Society guidelines for the screening of semen donors for donor insemination. Hum Reprod. 1999;14(7):1823–6.
50. The American Society for Reproductive Medicine. Guidelines for gamete and embryo donation. Fertil Steril. 1998;70(4 Suppl 3):1S–3

Chapter 7
Preservation of Sperm Isolates or Testicular Biopsy Samples

Bhushan K. Gangrade

The pioneering work of Steptoe and Edwards [1] in human reproduction leading to the birth of Louise Brown not only made history, but was also instrumental in developing a new field of reproductive technology for the treatment of infertility. While in vitro fertilization by coincubation of sperm and oocytes together works well when the sperm parameters (e.g., count, motility, morphology, etc.) are relatively within normal range, fertilization of oocytes is often compromised if the male partner has suboptimal semen characteristics. Microinjection-assisted fertilization techniques, such as zona pellucida drilling, subzona insemination, and direct sperm injection, have been historically employed to overcome the male factor infertility with varying success. The development of the technique of intracytoplasmic sperm injection (ICSI) by Palermo et al. [2] has revolutionized the treatment of male factor infertility. Theoretically, this technique (ICSI) requires one viable sperm per oocyte. Originally, ICSI was employed to treat male factor infertility using ejaculated sperm; however, soon thereafter, successful pregnancies were reported following the injection of freshly retrieved testicular [3, 4] and epididymal [5] sperm and to treat infertility secondary to nonobstructive azoospermia (NOA) using testicular sperm [6]. Attempts at the retrieval and cryopreservation of testicular sperm in men with azoospermia have become integral practice in most if not all assisted reproductive technology programs.

Azoospermia

The complete absence of sperm in the ejaculate is referred to as azoospermia. This condition is observed in approximately 1% of the general population and approximately 10% of infertile men [7, 8]. The initial observation of the absence of sperm

B.K. Gangrade, Ph.D. (✉)
IVF Laboratory, Center for Reproductive Medicine,
3435 Pinehurst Avenue, Orlando, FL 32804, USA
e-mail: bkgangrade@hotmail.com

E. Seli and A. Agarwal (eds.), *Fertility Preservation in Males: Emerging Technologies and Clinical Applications*, DOI 10.1007/978-1-4614-5620-9_7,
© Springer Science+Business Media New York 2012

in the liquefied ejaculate is confirmed by centrifugation (300×g for 15 min) of the semen followed by a complete systematic microscopic search of the resuspended pellet. Failure to observe any sperm in the ejaculate on two occasions following at least 2–3 days of abstinence is suggestive of the diagnosis of azoospermia. Absolute absence of sperm in the ejaculate may be caused by either an obstruction in the reproductive tract (obstructive azoospermia) or by actual failure of sperm production in the seminiferous tubules (NOA). Occasionally, in some NOA patients, small isolated portions of seminiferous tubules may show some spermatogenesis, but the number of sperm released is so few that no sperm are found in the ejaculate. Such patients are, therefore, essentially given a diagnosis of NOA.

Obstructive Azoospermia

Obstructive azoospermia may be congenital (congenital bilateral absence of the vas deferens or CBAVD) or acquired (vasectomy, infection in the reproductive tract, or surgery-related). Men with obstructive azoospermia usually have normal spermatogenesis and sperm may be directly retrieved surgically from the testis or epididymis.

Nonobstructive Azoospermia

Some of the common causes of the NOA are Klinefelter's syndrome, testicular failure secondary to radiation or chemotherapy, primary germ cell failure, Sertoli cell-only syndrome, or hypophyseal (hypogonadotropic) hypogonadism. Other causative factors may include testicular torsion, mumps orchitis, and cryptorchidism. Some men with NOA exhibit focal spermatogenesis in isolated seminiferous tubules. When less than six mature spermatids are observed in the cross section by histological analysis, usually sperm are absent in the ejaculate [9], but a thorough examination of testicular biopsy tissue may yield few sperm in isolated tubules that may be enough to fertilize occytes by ICSI and achieve pregnancy. A variation of the technique of open testicular biopsy (Testicular sperm extraction, TESE), sometimes referred to as micro-TESE, involves bisection of the testis and thorough examination of the seminiferous tubules by the urologist under optical magnification for the identification and excision of portions of seminiferous tubules with active spermatogenesis. Successful sperm retrieval following micro-TESE in almost 50% of men diagnosed with NOA has been reported [10, 11]. A comparison of standard TESE with micro-TESE shows that the latter surgical procedure is more effective in successful sperm retrieval in NOA patients [12]. A positive correlation between testis volume and the probability of retrieving sperm has been reported [13]. In men with small or atrophied testis (volume <5 mL), micro-TESE offers a better chance of retrieving sperm as compared with TESE. Furthermore, FSH levels in men with NOA negatively correlate with the probability of finding spermatozoa in testis; however, this correlation has recently been refuted by some researchers [13].

In NOA, in order to find sufficient sperm for ICSI, it may be necessary to remove relatively larger amount of testicular tissue. Sometimes, multiple biopsies at different loci are needed which in turn may negatively affect testicular androgen production. The decrease in androgen production following removal of testicular tissue may be more pronounced in men with relatively small or atrophied testis. Even though micro-TESE is a more invasive surgical procedure, the amount of tissue excised in micro-TESE is significantly less as only a few seminiferous tubules that show active spermatogenesis are removed.

In men with Klinefelter's syndrome, sperm can be retrieved in 21–69% of patients [13, 14]. Since these men have a high incidence of testicular atrophy, micro-TESE might be a better option for sperm retrieval. In order to avoid repeat testicular biopsy surgeries for future IVF-ICSI attempts, sperm may be frozen in multiple aliquots as described in a later section.

The role of Y chromosome in spermatogenesis is well-recognized. Microdeletions in the proximal regions of the long arm (AZFa and/or AZFb) of the Y chromosome have been shown to cause partial or complete failure of spermatogenesis. The extent of severity in terms of spermatogenesis depends on the size of the microdeletion in AZFa or AZFb region. In contrast, microdeletions in the distal AZFc region may cause mild to severe oligospermia and in some cases even azoospermia [15]. Approximately 10–15% of the patients of NOA may have microdeletions in one or more regions of the Y chromosome [16]. These couples if undergoing IVF-ICSI with testicular sperm should be advised extensively regarding the transmission of defective Y chromosome to the male progeny.

In men with NOA, even though it is possible to find sperm in isolated seminiferous tubules following testicular biopsy, the yield (of sperm) is significantly less than that usually retrieved in normospermic patients with obstruction. Testicular sperm from NOA patients can be frozen and used for ICSI at a later time. Several techniques for sperm retrieval from azoospermic men have been described. In men with obstructive azoospermia, sperm can be retrieved in almost all cases by testicular fine needle aspiration (TEFNA), testicular sperm aspiration (TESA), percutaneous epididymal sperm aspiration (PESA), or microepididymal sperm aspiration (MESA). In some cases, open conventional testicular biopsy or TESE may be preferred especially if the intent is to cryopreserve multiple aliquots of sperm for future use or other retrieval techniques have been unsuccessful in sperm retrieval.

Reasons for the Cryopreservation of Testicular Sperm

Cryopreservation of surgically retrieved testicular sperm from male partner prior to initiating the treatment cycle of the female almost ensures the availability of the male gamete on the day of the egg retrieval. This reason for cryopreservation may not be of as much significance in men with obstructive azoospermia since such patients usually have normal spermatogenesis. On the other hand, in patients with NOA, advance cryopreservation of sperm offers a psychological relief from the anxiety and uncertainty related to sperm retrieval on the same day when the female partner is undergoing oocyte retrieval. In addition, testicular sperm cryopreservation avoids the logistical difficulties of coordinating the surgeries for both the male and the female on the same day [17]. Administration of anesthesia to both the male and the female partner on the same day for gamete retrieval makes it necessary on patient's part to ask for outside help from friend or relative to accompany and transport them from the clinic/hospital after the surgery. This puts extra burden on the couple and also makes it difficult to keep the privacy, if they so desire, regarding their infertility treatment. Repeated testicular aspirations or biopsies have been shown to adversely affect testicular function [18], and cryopreservation of multiple aliquots of sperm avoids the need for repeated surgeries if the couple desires additional IVF/ICSI attempts.

Men who have had vasectomy, but decide to regain their fertility, often opt for vas reversal as the first line of treatment. Even though the vas patency rate for an obstructive interval of 10–15 years is quite high (approximately 74%), the reconstruction surgery is unsuccessful in approximately one quarter of the patients [19]. Between 50% and 80% of the patients who have undergone vasectomy reversal exhibit normal semen parameters (sperm concentration, morphology, motility, and progression) after the reconstruction surgery, but fail to achieve pregnancy due to the development of antisperm antibodies [20, 21]. Since the spermatozoa come into contact with the antibodies in the epididymis or in the ejaculatory duct, sperm retrieval from the site of origin, i.e., testis, avoids contact of spermatozoa with necrotic milieu. Cryopreservation of testicular/epididymal sperm in men at the time of vas reversal surgery avoids the need for a repeat surgical intervention if either vas reconstruction is unsuccessful or if the couple fails to achieve a successful pregnancy despite normal semen parameters. Use of cryopreserved testicular sperm has been documented in the case of necrospermia (a condition in which ejaculate contains only dead sperm) secondary to antisperm antibodies [22]. Exploratory or diagnostic testicular biopsies in men seeking infertility treatment if scheduled as "possible testicular sperm freeze" may in some cases avoid the need for another testicular biopsy [23]. Cryopreservation of testicular sperm has also been successfully used in men with spinal cord injury [24].

Cryopreservation of testicular sperm to preserve fertility in prepubertal boys undergoing gonadotoxic therapy for cancer has also been suggested [25, 26]. Even if sperm are not found in the ejaculate, there may be focal spermatogenesis in isolated tubules and enough testicular sperm may be retrieved and cryopreserved in prepubertal (10–13 years old) boys.

Retrieval of Sperm from Testicular Tissue

Testicular parenchymal biopsies are taken from a nonvascular region in or near equatorial plane. Approximately 5×5 mm biopsy samples from obstructive azoospermic men provide enough sperm that can be frozen in 3–4 (or more) aliquots, thus allowing several IVF-ICSI attempts. A small (1–2 mm) piece of seminiferous tubule in a drop of culture medium (HEPES-HTF) is transferred to a glass slide for a squash preparation and observed under the microscope and 200–400× magnification. It is usually not feasible to perform a quantitative sperm analysis in terms of the concentration, motility, vitality, and morphology in the testicular biopsy samples because of the extremely low sperm concentration. A subjective assessment of the number of spermatozoa, however, should be made to determine the number of aliquots that may be frozen. The testicular parenchyma is transferred to a petri dish and minced using a pair of fine-tipped sterile scissors or alternatively teased with a pair of 27-gauge needles attached to tuberculin syringes and squeezed to release the contents of the seminiferous tubules. A comparison of four different mechanical methods (shredding between two glass slides, mincing, vortexing, and crushing using electrical Potter) showed identical yields of sperm [27]. Since human sperm can survive at room temperature for prolonged duration [28], heated temperature-controlled (37°C) surface is not needed. Some investigators, however, prefer to maintain the temperature of contents at 37°C [29]. If the testicular cell suspension exhibits excessive red blood cells, short (5 min) exposure to erythrocyte lysis buffer (155 mM NH_4Cl, 10 mM $KHCO_3$, 2 mM EDTA; pH 7.2) effectively removes blood cells [27] without causing any detrimental effect to sperm [27, 30].

Isolation and Cryopreservation of Testicular Sperm

Thawing of Testicular Sperm

Frozen testicular sperm is thawed on the day of the oocyte retrieval approximately 3 h before performing ICSI. It is prudent to ascertain that mature (Metaphase II) oocytes are available for ICSI prior to thawing sperm. A vial of frozen sperm is removed from the liquid nitrogen and thawed at room temperature for 10 min. The contents of the vial are diluted by slowly adding equal volume of HSA (10 mg/mL)-supplemented HEPES-HTF medium. The suspension is transferred to a 15-mL conical bottom tube and centrifuged at $300 \times g$ for 10 min. The supernatant is discarded and the crude sperm pellet is washed once by resuspending in 1 ml HEPES-HTF and centrifuged again as above. The final sperm pellet is resuspended in 0.1 mL medium and stored at room temperature until ICSI.

Selection of Viable Sperm in Testicular Biopsy Samples

With the introduction of ICSI, now it is possible to achieve high fertilization rate with sperm irrespective of the source. Testicular sperm are as effective as ejaculated sperm provided that viable sperm are selected for injection [31]. Testicular sperm usually are immature and immotile and achieve motility during the passage through the epididymis. The main characteristic that easily differentiates viable sperm from nonviable is the motion or slight tail movement. There is a further reduction in sperm viability after freeze and thaw. Since the testicular sperm are largely immotile, it is important to identify the viable sperm for ICSI. Several methods have been proposed to select viable sperm from fresh or frozen testicular specimen. Prolonged in vitro culture of fresh [32–35] and frozen [36] testicular sperm for 48–72 h significantly improves the proportion of motile sperm in the sample. The in vitro culture appears to mimic the maturation of spermatozoa and acquisition of motility in the epididymis. In fact, fresh testicular sperm may survive for up to 2 weeks at 37°C, whereas at slightly lower temperature (32°C) that is in line with scrotal temperature, their survival may prolong even longer. In case of obstructive azoospermia, in vitro culture of testicular sperm for 72 h followed by cryopreservation offers a high rate of recovery of motile spermatozoa at thaw [36].

Identification of the physiological integrity of the plasma membrane has been used to differentiate between viable and nonviable spermatozoa for ICSI. Live sperm when exposed to a hypo-osmotic solution (75 mmol/L fructose and 25 mmol/L sodium citrate dehydrate) exhibit swelling/curling of the tail, whereas dead sperm remain unaffected [37]. Successful pregnancies were achieved by incorporating hypo-osmotic swelling (HOS) test as the method to differentiate and select viable sperm for ICSI. A comparison of three different hypo-osmotic solutions (original Jeyendran solution containing fructose and sodium citrate as above, a mixture of 50% culture medium and 50% MilliQ water, and MilliQ water) showed that 50% culture medium with 50% MilliQ water offers better sperm quality than original HOS solution or pure MilliQ water [38]. Use of hypo-osmotic solution consisting of culture medium diluted (1:1) with MilliQ water is a simple and effective method for the selection of viable testicular and ejaculated spermatozoa [39].

Pentoxifylline, an inhibitor of phosphodiesterase activity, enhances sperm motility by increasing the accumulation of intracellular cAMP. In vitro exposure to pentoxifylline also causes an increase in acrosome reaction [40], sperm penetration in zona-free hamster oocytes [41], and preserves functional membrane integrity of human spermatozoa [42]. Pentoxifylline has been shown to enhance motility in normozoospermic as well as asthenozoospermic samples [43]. In vitro exposure of sperm to pentoxifylline prior to intrauterine insemination significantly increases pregnancy rate as compared to standard sperm preparation [44, 45]. The inability of pentoxifylline to improve in vitro fertilization in patients with history of failed or reduced fertilization [46, 47] is of limited significance in the era of ICSI. There is no doubt that pentoxifylline treatment significantly improves sperm motility in vitro and can be used to select viable but otherwise nonmotile testicular sperm. The average time to search for viable (motile) sperm following pentoxifylline exposure in fresh and frozen azoospermic TESE samples is significantly less than in its absence [48].

For induction of motility in testicular (fresh or frozen) sperm, a 30 μL flattened drop of pentoxifylline (3.6 mM) is made in the ICSI dish. In addition, the ICSI dish also contains drops of HEPES-HTF for holding oocytes and a drop of polyvinylpyrrolidone (PVP; 10%) for immobilizing the sperm. For ICSI, 5 μL fresh or thawed testicular sperm sample is added to the pentoxifylline drop. Motility is induced in testicular sperm in approximately 10 min. The motile hyperactivated sperm swim to the periphery of the pentoxifylline drop and are picked up using the ICSI pipette and transferred to the PVP drop for washing and immobilization. ICSI is then performed according to the standard protocol. The direct addition of sperm to the pentoxifylline drop is a simple and quick procedure to select viable but otherwise nonmotile sperm for ICSI.

Cryopreservation of Testicular Sperm Inside Evacuated Zona Pellucida

Cryopreservation of testicular sperm as described earlier works quite well (such as in case of obstructive azoospermia) when several hundred to a few thousand sperm are available for freezing. There is always a decrease in the proportion of viable spermatozoa after thawing the frozen testicular preparation. This is usually not of much concern in frozen-thawed obstructive azoospermic TESE samples since sufficient numbers of viable spermatozoa are present for ICSI. However, in patients with NOA, finding sperm in post-thaw TESE sample at the time of ICSI could be challenging and frustrating. In our experience, in frozen-thawed TESE samples from NOA patients, it may take up to 30 min or more to find a single twitching sperm suitable for ICSI. The prolonged exposure of oocytes to external environment while searching for a viable spermatozoon compromises the embryo quality and may adversely affect the outcome of the procedure.

In 1997, Cohen et al. [49] reported a unique method of cryopreserving a single sperm or few spermatozoa within an evacuated zona from hamster or mouse egg. This method can be used to freeze sperm in extreme cases when less than a few hundred sperm are available. This approach requires extensive preparation, expertise, and time and has been sparingly utilized in clinical setting. The empty zona envelop serves as a small container for spermatozoa that is easy to locate and handle. Locating the sperm inside the zona is easy and relatively less time consuming. It almost guarantees the availability of sperm at the time of ICSI. A brief description of the method of sperm cryopreservation inside evacuated zona is presented below.

If TESE reveals the presence of sperm in the sample, a 50-mm dish (Falcon 35–1006) is used to isolate and retrieve spermatozoa from the crude TESE preparation. Incubation of TESE preparation at room temperature for several (up to 24) hours may induce in vitro maturation and motility in viable sperm. A large, surface flat drop (called sedimentation layer) of HEPES-HTF with HSA (10 mg/mL) is made in the upper center portion of the dish. Approximately 5 μL of testicular sample preparation is added to the center of the sedimentation layer carefully without disturbing the contents. The cellular debris falls to the bottom in the sedimentation layer while rare sperm slowly swim to outer peripheral edges.

Frozen hamster oocytes purchased from a commercial vendor (Charles River Inc., Wilmington, MA, USA) are emptied of the ooplasm as follows. Cryostraws containing eggs are thawed as per supplier's instructions. Morphologically healthy oocytes are washed in the culture medium and transferred to 5 μL droplet of HEPES-HTF with HSA (10 mg/mL) in a micromanipulation dish (Falcon, 35–1006). The drop is covered with mineral oil to prevent evaporation. The oocyte is held against the holding pipette and penetrated with an ICSI-like micropipette (15 μm). The ooplasm is aspirated by moving the micropipette around the egg and applying suction until the zona is empty. Once the zona envelop has been cleared of ooplasm, spermatozoa in sedimentation drop are aspirated using an ICSI pipette and transferred to a 5 μL droplet of PVP (10%). Depending on the total number of sperm available for freezing, one to five sperm are loaded into the ICSI pipette and injected in the empty zona. The spermatozoa are not immobilized at this time. The zona containing sperm is then loaded into a cryostraw (one zona per straw) and frozen according to laboratory's standard sperm cryopreservation procedure.

In order to recover the sperm on the day of ICSI, straws are thawed by immersing in a water bath at 30°C. The contents of the straw are expelled into a drop in a culture dish. The zona is retrieved, washed in HTF-HEPES medium, and transferred to an ICSI dish. The ICSI dish contains a PVP drop surrounded by several drops of HTF-HEPES. The zona envelop is held onto a holding pipette and an ICSI micropipette preloaded with 10% PVP solution is inserted into the zona. It is preferable to enter the zona through the original hole that was made to deposit the sperm inside. The sperm are aspirated and transferred to the PVP drop, immobilized, and used for ICSI. The HTF-HEPES drops surrounding the PVP drop are used to contain cumulus-free oocytes.

Successful pregnancies and live birth were reported using post-thawed spermatozoa frozen inside hamster zonae [50]. Modifications of the technique of sperm cryopreservation in evacuated zona have been reported and claimed to be superior to conventional freezing methods [51, 52]; however, the feasibility of its widespread use and acceptance in clinical context remain to be seen.

Use of cryoloop (Hampton Research, Laguana, CA) to cryopreserve single (or few) sperm has also been attempted [53], but has not gained wide acceptance for clinical use.

Cryopreservation of Intact Seminiferous Tubules and Testicular Tissue

Cryopreservation of excess testicular sperm after ICSI has been a common practice in IVF laboratories. Some health care providers and/or patients may prefer to cryopreserve testicular sperm before the retrieval. Cryopreservation of intact testicular tissue on the other hand has achieved little attention and acceptance. There are a few reports describing the feasibility of testicular tissue cryopreservation in obstructive and NOA. One of the earlier attempts to cryopreserve testicular tissue involved freezing of isolated seminiferous tubules from men with obstructive and NOA [54]. The freezing protocol for tubule cryopreservation is presented below.

Intact tubules obtained by PESA are frozen in sperm-freezing medium containing 15% glycerol. Briefly, pieces of seminiferous tubules are washed in culture medium and transferred to cryovials in 0.5 mL of HTF-HEPES. Equal volume (0.5 mL) of sperm-freezing medium is added slowly to the vial and allowed to equilibrate for 30 min at room temperature. Cryovials are placed in a −20°C standard freezer (30 min), exposed to the liquid nitrogen vapor (30 min) in a wire basket, and then plunged in liquid nitrogen for storage.

The seminiferous tubules are thawed at room temperature for 15 min and then washed in HTF-HEPES medium. Spermatozoa are squeezed out of the tubules by 27-gauze tuberculin needle, washed by centrifugation, and used for ICSI.

Cryopreservation of testicular tissue from men followed by successful extraction of spermatozoa [55] is also reported. This method of freezing tissue is similar to that used for cryopreservation of seminiferous tubules described above. The biopsy tissue is incubated for 30 min in sperm-freeze medium and then slow cooled for 10 min at 4°C. The cryovials are then vapor cooled by hanging in liquid nitrogen vapor for 30 min and plunged in liquid nitrogen.

The thawing procedure involves warming the tissue in a water bath (37°C). The tissue pieces are washed with warm (37°C) HEPES-HTF and minced in a petri dish using 27-gauze syringe needles or a pair of sharp scissors. The testicular tissue suspension is then processed as described earlier for isolating sperm and performing ICSI.

The reason for the lack of interest on the part of reproductive laboratory personnel in freezing the testicular tissue appears to be the extra time and effort needed for preparation of sperm on the day of ICSI. In comparison, thawing of previously frozen testicular sperm is a relatively simple and less time-consuming procedure. Moreover, there is no additional advantage of freezing the testicular tissue when compared with freezing the sperm in azoospermic men.

Successful freezing and thawing of testicular tissue in prepubertal boys, who lack sperm in the ejaculate but require chemo- or radiation therapy for cancer treatment, would open a major avenue in fertility preservation [56]. It has been proposed [57] that cryopreservation of testicular cell suspension may be a better option than tissue cryopreservation in young boys who have not yet achieved spermatogenesis. It is, however, important to note that the basic architecture of testicular tissue is important in the development of spermatozoa from spermatogonial cells since physical contact with other cells (e.g., Sertoli cells) is required for spermatogenesis [58]. Keeping this in view, attempts have been made to freeze testicular tissue from young prepubertal males undergoing gonadotoxic therapy. Keros et al. [56] reported successful cryopreservation of prepubertal testicular tissue in boys undergoing cancer therapy. Small (2–5 mm^3) pieces of testicular tissue were frozen using 5% dimethyl sulfoxide (DMSO) and a slow freezing protocol. The tissue was cooled at a rate of −1°C/min from room temperature to 0°C and held for 5 min followed by cooling at −0.5°C/min until temperature dropped to −8°C. Manual seeding was performed at this time and cryovials were held at this temperature for 10 min. The cooling ramp then continued at the rate of −5°C/min until −40°C, held for 10 min, and the freezing continued to −70°C at −7°C/min. Thereafter, tissue vials were plunged into liquid nitrogen.

Thawing of testicular tissue was carried out by immersing the vials in a 37°C water bath until the ice melted. Tissue pieces were washed in culture medium and analyzed for structural changes by light and electron microscopy and also by immunohistochemistry. A comparison of fresh and frozen prepubertal testicular sample tissue revealed that frozen-thawed tissue maintained normal structure. Frozen tissue also exhibited good survival of spermatogonia in seminiferous tubules. The cell-to-cell contact and attachment of spermatogonia to the basal membrane was comparable to that noted in the fresh biopsy tissue. In this study, no attempt was made to analyze the endocrine status and responsiveness (i.e., androgen production by Leydig cells in response to gonadotropins) of the frozen-thawed sample. Further studies are required to explore and develop techniques and assess the feasibility of the whole tissue (testicular) freezing for preservation of fertility.

Posthumous Reproduction

Posthumous reproduction refers to the birth of a child after either one or both the genetic parents have died. Needless to say that posthumous conception and reproduction by IVF is a topic of ethics and law and is beyond the scope of this chapter and the expertise of the author. A brief discussion on this topic, however, deserves a place since now it is technically possible to cryopreserve both the sperm and eggs and to achieve conception by using the cryostored gametes of the dead person. Cryopreservation of oocytes for autologous use is still considered experimental and is rarely practiced. In contrast, sperm cryopreservation has been in practice for decades. Intentional cryopreservation of sperm by consenting male is a standard practice, where the intent of the male is clear, i.e., to achieve pregnancy with the spouse or female partner. Nowadays, some reproductive clinics require the male partner to record his wish regarding the fate of his sperm in the event of his death. In contrast, the practice of harvesting sperm from a deceased man (posthumous procurement of sperm) [59], usually at the request of his female partner, is quite controversial. It is obvious that in almost all such instances there is no directive from the deceased regarding the use of his sperm upon his death. In USA, the existing law does not provide clear guidance to whether the sperm may be retrieved form a dead man. The Ethics committee of the American Society for Reproductive Medicine issued a report [60] in September 2004 recognizing that "posthumous reproduction will be employed in instances when a couple faced with the imminent death of a partner or in anticipation of radiation or chemotherapy for cancer will ask to have gametes obtained and stored. Should death occur, posthumous reproduction using the stored gametes may be requested by the surviving partner." The Ethics committee further stated that "it is the responsibility of the health care provider (Reproductive Endocrinologist in this instance) to ascertain that appropriate informed consents are obtained and to ensure adequate screening and counseling of all concerned parties."

There are significant differences in beliefs and laws on posthumous reproduction depending on the country and religion. In Israel, women can have the sperm harvested and cryopreserved from their dead husbands even if the husband had not given any prior consent. Only in case the husband has clearly made his wish that he is not willing his sperm to be used for insemination after his death, the sperm cannot be retrieved postmortem on wife's request. The same guidelines apply to life partners (even if not married), but bar parents and any other relative to request sperm preservation from the dead man. In Japan, posthumous reproduction has been practiced several times in the absence of any legislative regulation. A survey of over 3,000 Japanese people revealed that there is strong support from the public in favor of posthumous conception [61]. In Belgium, posthumous reproduction is permitted. In several eastern European countries like Cyprus, the Czech Republic, Latvia, Lithuania, Malta, Poland, and Slovakia, there is no legislation concerning posthumous reproduction. Posthumous sperm retrieval is prohibited in Hungary, Slovenia, and Malaysia [62]. Australia (State of Victoria), Canada, Germany, and Sweden prohibit posthumous sperm procurement [63] while in France posthumous insemination is prohibited [64]. In terms of religion, posthumous procurement of sperm (or oocytes) is strictly prohibited in Islam [65].

Conclusions

Retrieval and cryopreservation of testicular sperm has become a cornerstone of infertility treatment in azoospermic men. In obstructive azoospermia, sperm may be retrieved from testis in almost 100% patients and can be frozen in multiple aliquots to avoid repeated surgeries. In approximately 50% of patients with NOA, enough sperm may be retrieved for IVF-ICSI. Any excess sperm remaining after ICSI can be frozen for future IVF attempts. Frozen testicular sperm are as effective in achieving successful pregnancies as freshly isolated testicular or ejaculated sperm. Treatment with motility-enhancing agents, such as pentoxifylline, induces motility in frozen-thawed testicular sperm and facilitates selection of viable sperm for ICSI. Special techniques, such as cryopreservation of single sperm within evacuated hamster zona or on a cryoloop, allow storage of extremely poor sperm samples. In USA, at this time, there is no legislative directive with respect to the posthumous procurement of sperm.

See Chap. 13

References

1. Steptoe PC, Edwards RG. Birth after the reimplantation of a human embryo. Lancet. 1978;2:366.
2. Palermo G, Joris H, Devroey P, et al. Pregnancies after intracytoplasmic sperm injection of single spermatozoon into an oocyte. Lancet. 1992;340:17–8.

3. Schoysman R, Vanderzwalmen P, Nijs M, et al. Pregnancy after fertilization with human testicular spermatozoa. Lancet. 1993;342:1237.

4. Craft I, Bennett V, Nicholson N. Fertilizing ability of testicular spermatozoa. Lancet. 1993;342:864.

5. Tournaye H, Devroey P, Liu J, et al. Microsurgical epididymal sperm aspiration and intracytoplasmic sperm injection: a new effective approach to infertility as a result of congenital absence of the vas deferens. Fertil Steril. 1994;61:1045–51.

6. Devroey P, Liu J, Nagy ZP, et al. Pregnancies after testicular sperm extraction and intracytoplasmic sperm injection in non-obstructive azoospermia. Hum Reprod. 1995;10:1457–60.

7. Willott GM. Frequency of azoospermia. Forensic Sci Int. 1982;20:9–10.

8. Jarow JP, Espeland MA, Lipschultz LI. Evaluation of the azoospermic patients. J Urol. 1989;142:62.

9. Silber SJ. Microsurgical TESE and distribution of spermatogenesis in non-obstructive azoospermia. Hum Reprod. 2000;15:2278–84.

10. Tallarini A, Borini A, Bonu MA, et al. Testicular fine needle aspiration in non-obstructive azoospermia. Fert Steril. 2002;78(Suppl):S209.

11. Ishikawa T, Nose R, Yamaguchi K, et al. Learning curves of microdissection of testicular sperm extraction for nonobstructive azoospermia. Fert Steril. 2010;94:1008–11.

12. Colpi GM, Colpi EM, Piediferro G. Microsurgical TESE versus conventional TESE for ICSI in non-obstructive azoospermia: a randomized controlled study. Reprod Biomed Online. 2009;18:315–9.

13. Turunc T, Gul U, Haydardedeoglu B, et al. Conventional testicular sperm extraction combined with the microdissection technique in nonobstructive azoospermic patients: a prospective comparative study. Fertil Steril. 2010;94(6):2157–60.

14. Schiff JD, Palermo GD, Veeck LL, et al. Success of testicular sperm extraction and intracytoplasmic sperm injection in men with Klinefelter's syndrome. J Clin Endocrinol Metab. 2005;90:6263–7.

15. O'Brien KL, Varghese AC, Agarwal A. The genetic cause of male factor infertility: a review. Fertil Steril. 2010;93:1–12.

16. Pryor JL, Kent-First M, Muallem A, et al. Microdeletions in the Y chromosome of infertile men. New Engl J Med. 1997;336:534–40.

17. Ben-Yosef D, Yogev L, Hauser R, Yavetz I, et al. Testicular sperm retrieval and cryopreservation prior to initiating ovarian stimulation as the first line approach in patients with non-obstructive azoospermia. Hum Reprod. 1999;14:1794–801.

18. Schlegel P. Physiological consequences of TESE. Hum Reprod. 1996;11:159.

19. Kolettis PN, Sabanegh ES, D'Amico AM, et al. Outcome for vasectomy reversal performed after obstructive intervals of at least 10 years. Urology. 2002;60:885–9.

20. Kay DJ, Clifton V, Taylor JS, et al. Anti-sperm antibodies and sperm profiles in re-anastomosed men. Reprod Fertil Dev. 1993;5:135–9.

21. Royle MG, Parslow JM, Kingscott MM, et al. Reversal vasectomy: the effects of sperm antibodies on subsequent fertility. Br J Urol. 1981;53:654–9.

22. Chavez-Badiola A, Drakeley AJ, Finney V, et al. Necrospermia, antisperm antibodies, and vasectomy. Fertil Steril. 2008;89:723.e5–e7.

23. Verheyen G, Vernaeve V, Van Landuyt L, et al. Should diagnostic testicular sperm retrieval followed by cryopreservation for later ICSI be the procedure of choice for all patients with non-obstructive azoospermia. Hum Reprod. 2004;19:2822–30.

24. Kanto S, Uto H, Mayumi T, et al. Fresh testicular sperm retrieved from men with spinal cord injury retains equal fecundity to that from men with obstructive azoospermia via intracytoplasmic sperm injection. Fertil Steril. 2004;92:1333–6.

25. Hovatta O. Cryopreservation of testicular tissue in young cancer patients. Hum Reprod Update. 2001;7:378–83.

26. Keros V, Hultenby K, Borhstrom B, et al. Method of cryopreservation of testicular tissue with viable spermatogonia in pre-pubertal boys undergoing gonadotoxic canter treatment. Hum Reprod. 2007;22:1384–95.

27. Verheyen G, De Croo I, Tournaye H, et al. Comparison of four mechanical methods to retrieve spermatozoa from testicular tissue. Hum Reprod. 1995;10:2956–9.
28. Petrella C, Hsieh J, Blake E, et al. Human sperm can survive at room temperature for weeks: measure by motility and viability of sperm maintained under various conditions. Fertil Steril. 2003;80(Suppl):S210.
29. Habermann H, Seo R, Cislak J, et al. In vitro fertilization outcomes after intracytoplasmic sperm injection with fresh or frozen-thawed testicular spermatozoa. Fertil Steril. 2000;73:955–60.
30. Nagy ZP, Verheyen G, Tournaye H, et al. An improved treatment procedure for testicular biopsy specimens offers more efficient sperm recovery: case series. Fertil Steril. 1997;68:376–9.
31. Palermo GD, Schlegel PN, Hariprashad J, et al. Fertilization and pregnancy outcome with intracytoplasmic sperm injection for azoospermic men. Hum Reprod. 1999;14:741–8.
32. Craft I, Tsirigotis M, Zhu J. In vitro culture of testicular sperm. Lancet. 1995;346:1438.
33. Edirisinghe WR, Junk SM, Matson PL, et al. Case report: changes in motility pattern during in vitro culture of fresh and frozen thawed testicular and epididymal spermatozoa: implications for planning treatment by intracytoplasmic sperm injection. Hum Reprod. 1996;11:2474–6.
34. Liu J, Garcia E, Baramki A. The difference in outcome of in vitro culture of human testicular spermatozoa between obstructive and non-obstructive azoospermia. Hum Reprod. 1996;11:1587–8.
35. Zhu J, Tsirigotis M, Pelekanos M, et al. In vitro maturation of human testicular spermatozoa. Hum Reprod. 1996;11:231–2.
36. Emiliani S, Van den Bergh M, Vannin AS, et al. Increased sperm motility after in-vitro culture of testicular biopsies from obstructive azoospermic patients results in better post-thaw recovery rate. Hum Reprod. 2000;15:2371–4.
37. Jeyendran RS, Van der Ven HH, Perez-Pelaez M, Crabo BG, Zaneveld LJD, et al. Development of an assay to assess the functional integrity of the human sperm plasma membrane and its relationship to other semen characteristics. J Reprod Fertil. 1984;70:219–28.
38. Verheyen G, Joris H, Crits K, et al. Comparison of different hypo-osmotic swelling solutions to select viable immotile spermatozoa for potential use in intracytoplasmic sperm injection. Hum Reprod. 1997;3:195–203.
39. Sallam HN, Farrag A, Agameya AF, et al. The use of a modified hypo-osmotic swelling test for the selection of viable ejaculated and testicular immotile spermatozoa in ICSI. Hum Reprod. 2001;16:272–6.
40. Cummins JM, Pember SM, Jequier AM, et al. A test of the human sperm acrosome reaction following ionophore challenge (ARIC)-Relationship to fertility and other semen parameters. J Androl. 1991;12:98–103.
41. Lambert HL, Steinleitner A, Eiserman J, et al. Enhanced gamete interaction in the sperm penetration assay after coincubation with pentoxifylline and human follicular fluid. Fertil Steril. 1992;58:1205–8.
42. Stanic P, Sonicki Z, Suchanek E. Effect of pentoxifylline on motility and membrane integrity of cryopreserved human spermatozoa. Int J Androl. 2002;25:186–90.
43. McKinney KA, Lewis SEM, Thompson W. Persistent effects of pentoxifylline on human sperm motility after drug removal in normozoospermic and asthenozoospermic individuals. Andrologia. 1994;26:235–40.
44. Negri P, Grechi E, Tomasi A, et al. Effectiveness of pentoxifylline in semen preparation for intrauterine insemination. Hum Reprod. 1996;6:1236–9.
45. Mehrannia T. The effect of pentoxifylline in semen preparation for intrauterine insemination. Pak J Med Sci. 2009;25:359–63.
46. Tournaye H, Janssens R, Camus M, et al. Pentoxifylline is not useful in enhancing sperm function in cases with previous in-vitro fertilization failure. Fertil Steril. 1993;59:210–5.
47. Tournaye H, Janssens R, Verheyen G, et al. An indiscriminate use of pentoxifylline does not improve invitro fertilization in poor fertilizers. Hum Reprod. 1994;9:1289–92.
48. Griveau JF, Lobel B, Laurent MC, et al. Interest of pentoxifylline in ICSI with frozen-thawed testicular spermatozoa from patients with non-obstructive azoospermia. Reprod Biomed Online. 2007;12:14–8.

49. Cohen J, Garrisi GJ, Congedo-Ferrara TA, et al. Cryopreservation of single human spermatozoa. Hum Reprod. 1997;12:994–1001.
50. Walmsley R, Cohen J, Ferrara-Congedo T, et al. The first births and ongoing pregnancies associated with sperm cryopreservation within evacuated egg zonae. Hum Reprod. 1998;13:61–70.
51. Hsieh YY, Tsai HD, Chang CC, Lo HY. Cryopreservation of human spermatozoa within human or mouse empty zona pellucidae. Fertil Steril. 2000;73:694–8.
52. Ye Y, Xu C, Qian Y, et al. Evaluation of human sperm function after being cryopreserved within the zona pellucidae. Fertil Steril. 2009;92:1002–8.
53. Desai NN, Blackmon H, Goldfarb J. Single sperm cryopreservation on cryoloops: an alternative to hamster zona for freezing individual spermatozoa. Reprod Biomed Online. 2004;9:47–53.
54. Allan JA, Cotman AS. A new method for freezing testicular biopsy sperm: three pregnancies with sperm extracted from cryopreserved sections of seminiferous tubules. Fertil Steril. 1997;68:741–4.
55. Scholtes MCW, van Hoogstraten DG, Schmoutziguer A, et al. Extraction of testicular sperm from previously cryopreserved tissue in couples with or without transport of oocytes and testicular tissue. Fertil Steril. 1999;72:785–91.
56. Keros V, Hultenby K, Borgstrom B, et al. Methods of cryopreservation of testicular tissue with viable spermatogonia in pre-pubertal boys undergoing gonadotoxic cancer treatment. Hum Reprod. 2007;22:1384–95.
57. Brook PF, Radford JA, Shalet SM, et al. Isolation of germ cells from human testicular tissue for low temperature storage and autotransplantation. Fertil Steril. 2001;75:269–74.
58. Ehmcke J, Hubner K, Scholer HR, et al. Spermatogonia: origin, physiology and prospects for conservation and manipulation of the male germ line. Reprod Fertil Dev. 2006;18:7–12.
59. Batzer FR, Hurwitz JM, Caplan A. Postmortem parenthood and the need for a protocol with posthumous sperm procurement. Fertil Steril. 2003;79:1263–9.
60. The Ethics committee of the American Society for Reproductive Medicine. Posthumous reproduction. Fertil Steril. 2004;82(Suppl):S260–2.
61. Ueda N, Kushi N, Nakatsuka M, et al. Study of views on posthumous reproduction, focusing on its relation with views on family and religion in modern Japan. Acta Med Okayama. 2008;62:285–96.
62. Dostal J, Utrata R, Loyka S, et al. Post-mortem sperm retrieval in new European union countries: case report. Hum Reprod. 2005;20:2359–61.
63. Bahadur G. Death and conception. Hum Reprod. 2002;17:2769–75.
64. Lansac J. French law concerning medically-assisted reproduction. Hum Reprod. 1996;11:1843–7.
65. Samani RO, Ashrafi M, Alizadeh L, et al. Posthumous assisted reproduction from Islamic perspective. Int J Fertil Steril. 2008;2:96–100.

Chapter 8
Germ Cell Transplantation and Neospermatogenesis

Queenie V. Neri, Zev Rosenwaks, and Gianpiero D. Palermo

Among the general population, the ability of man to procreate appears to have decreased progressively during the past half century [1]. An estimated 15% of couples in their reproductive age are afflicted by a range of causes pertaining to their ability to procreate, and about half of all infertility cases are directly attributed to the male partner.

Approximately 6% of males between the ages of 15 and 44 are deemed infertile or have their fecundity severely compromised [2]. Evaluation of endocrine data implies that the main causes of male infertility are hormonal disturbances and aberrations in the production of semen [1]. Absence of spermatozoa in the ejaculate is termed azoospermia. About 50% of all azoospermic cases are due to hypospermatogenesis while complete germ cell aplasia accounts for about 25%, the remainder are due to spermatogenic arrest [3].

Intracytoplasmic sperm injection (ICSI), focused at selecting a single viable spermatozoon to inseminate a mature oocyte, is now considered the best procedure to generate consistent fertilization and satisfactory pregnancy outcome [4, 5]. The ability of virtually any spermatozoon, regardless of its morphology or functional characteristics to induce oocyte activation and consequent pronuclear development has meant that even some azoospermic patients can be treated successfully with this procedure. Yet, in spite of these accomplishments in treating many forms of male factor infertility, there are still limitations since the fully developed spermatozoon remains the *conditio sine qua non* for treatment of male infertility, leaving those men afflicted by Sertoli-cell-only (SCO) syndrome or spermatogenic arrest as currently untreatable.

G.D. Palermo, M.D., Ph.D. (✉)
The Ronald O. Perelman and Claudia Cohen Center for Reproductive Medicine,
Weill Cornell Medical College, 1305 York Avenue, 6th Floor,
New York, NY 10021, USA

E. Seli and A. Agarwal (eds.), *Fertility Preservation in Males: Emerging Technologies and Clinical Applications*, DOI 10.1007/978-1-4614-5620-9_8,
© Springer Science+Business Media New York 2012

Recent research has aimed at the in vitro proliferation and maturation of germ cells, at least in animals [6]. However, spermatogonial stem cells (SSCs) are the only cell type in the male gonad that retain the ability to selfrenew and under certain stimuli to be driven into meiotic and post-meiotic differentiation [7]. Thus, SSCs are unique in that respect – they represent the progenitor cells leading to spermatozoa production, while at the same time displaying the ability to revert to a more totipotent state similar to the embryonic stem cell in order to reclaim their precursor state [8]. For this reason, mastering ability to culture and propagate SSCs in vitro may pave the way to an array of treatments aimed at alleviating several disorders characterizing male infertility.

Antineoplastic Treatment

Adults with Cancer

In 2009, approximately 62,000 individuals in the US between the ages of 20–39 years are expected to be diagnosed with cancer [9]. Many of them will require chemotherapy and/or radiotherapy regimens, antineoplastic, and cytotoxic agents that being highly gonadotoxic compromise the ability to procreate. Gonadal failure is one of the major consequences of cancer therapy, remarkably so after ionizing radiation. Many surveys on cancer survivors have found that they often have an increased risk of emotional distress related to infertility as a result of their treatment [10].

Depending on the underlying disease, the age of the oncological patient, the type of therapeutic agent used to treat the cancer, the cumulative radiation doses used or the duration of the chemotherapy, between 10% and 100% of surviving cancer patients will show impaired semen parameters following treatment [11]. It is estimated that between 15% and 30% of these cured cancer patients become permanently sterile [12]. There is the concern that many patients undergoing potentially sterilizing antineoplastic cancer treatments have not been appropriately counseled about their future fertility. Even though assisted reproductive technologies (ART) can successfully restore the reproductive capacity of many of these cancer patients, oncologists and their practitioners are still not fully aware of its limitations [13–15]. In fact, this lack of awareness in the medical community seems to be the main culprit for causing untimely cryostorage of gametes in these patients.

In the ICSI era, any semen sample, even those comprising only a few motile spermatozoa, may be considered for sperm cryopreservation, and this should be performed prior to any cytotoxic treatment. Since most patients will develop azoospermia 2 or 3 months following chemotherapy [12], cryobanking semen during the first month should still be advocated to overcome this major collateral effect [16]. However, patients should be informed that chemotherapy may induce genetic and epigenetic abnormalities in both the short- [17] and long-term [18, 19].

Azoospermia can sometimes be encountered in patients during the diagnosis of testicular cancer – this as a direct result of factors related to the malignancy on spermatogenesis [20]. At orchiectomy, these patients may be offered sperm harvesting by vasal, epididymal aspiration [21], or by testicular sperm extraction (TESE) [12] in order to bank their gametes prior to chemotherapy.

At our Center, the efficacy of standard IVF and ICSI has been evaluated in 118 couples who cryobanked semen before cancer treatment and then underwent 169 ART cycles [22]. These included men with cancer of the testis, prostate, brain, lung, pancreas, and bladder, as well as leukemia, lymphomas, sarcomas, and multiple myelomas. The mean age of the females at the time of their ART procedure was 34.8 ± 4 years and their husbands were 38.5 ± 8 years. The postthaw semen analysis had an average concentration of 40.9×10^6/mL and motility of 14.2%. The overall clinical pregnancy rate was 56.8% (96/169) with 11 (5.6%) undergoing spontaneous abortions while the remainder went to term delivery (85/169). Patients with prostate cancer displayed the worst semen parameters prior to banking and the lowest clinical pregnancy rates. In a subanalysis that compared the clinical pregnancy according to the insemination method, it appears that ICSI proved to be more efficacious than standard in vitro insemination ($P < 0.001$). Overall, ARTs particularly with ICSI is recommended for couples whose male partners are diagnosed with cancer and cryobanked their specimen before, after, and even during antineoplastic treatment.

Prepubertal Age Cancer

About one in every 600 children will develop cancer before the age of 15. However, as a result of better treatment options, today the cancer death rate in infants and children has decreased more than that for any other age group, with up to 75% of all cases now being curable. At present, one in 1,000 adults in the second or third decade of life is a childhood cancer survivor [23], and it has been estimated that this proportion will rise to one in 250 [24].

With childhood cancer being treated more successfully, the focus of treatment has shifted toward the quality of life after treatment, and one aspect commonly overlooked is the prevention of sterility. In adults and adolescents, semen banking or cryopreservation of testicular tissue before radiation/chemotherapy are valuable preventive measures that circumvent sterility after treatment. However, no such prevention is possible prior to puberty since no active spermatogenesis has occurred, and so an average of 30% of children cured of cancer remain sterile in the long term [12].

One possible remedy consists of SSC transplantation, the procedure in which either testis germ cells or stem cell colonies are transferred from a fertile donor into the seminiferous tubules of an infertile recipient. This is a technique currently being perfected in mice before attempts to apply it to man.

The first experimental attempt at autologous spermatogonial transplantation was conducted by Brinster and Avarbock in 1994 when mouse testicular cells obtained after enzymatic digestion were directly injected into seminiferous tubules of a recipient mouse previously made sterile with injections of busulfan (a strong alkylating agent) [25]. A month after post-germ cell transplantation, nascent germ cell

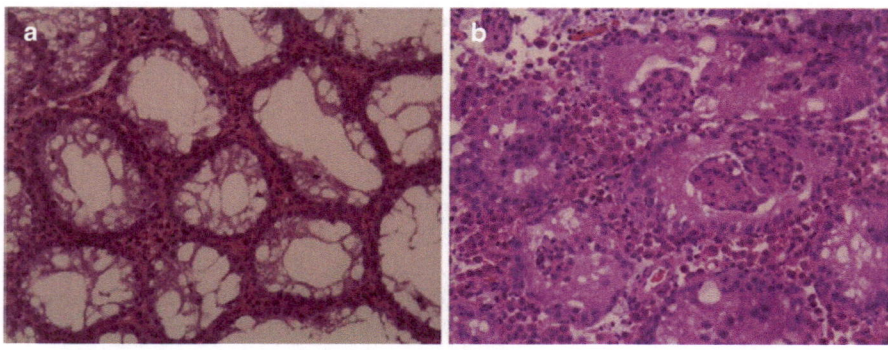

Fig. 8.1 Seminiferous tubules of SCID mouse (**a**) 4 weeks after busulfan treatment and (**b**) 1 week after injection of human spermatogenetic cells

colonies were identified as chains of cells on the tubular basement membrane. These spermatogonia stem cells (SSCs) implanted in the tubules then differentiated into later spermatogenic stages. In a further experiment, higher colonization rates of germ cells were reported by enhancing the transplanted donor cell population [26].

Following these demonstrations, xenogeneic germ cell transplantation was conducted with testicular cells from fertile transgenic rats injected into the seminiferous tubules of immunosuppressed mice [27]. In this case, mouse sertoli cells clearly were able to serve as supporting somatic cells in rat germ cell recolonization and eventual differentiation. Other species used for donor cell injection into mouse seminiferous tubules that achieved successful colonization and differentiation were hamster [28], rabbits, and dogs [29].

The first xenogeneic germ cell transplantation in which human testicular cells were injected into mouse seminiferous tubules was performed by our group in 2000 [30] that resulted in a lack of recolonization of SSCs in the recipient tubules of SCID or W/W^v mice after 150 days (Fig. 8.1). We have postulated then that lack of success may have been due to inter-species variability, to noncompatible cell adhesion molecules and/or to immunological rejection. Nagano et al. transplanted baboon germ cells into mice and found that germ cells were able to repopulate and survive for 6 months [31].

Investigators have reported the survival of at least some undifferentiated spermatogonia during distant xeno-germ cell transplantations [30–32]. In Reis et al., however, we did not find evidence of donor tissue survival following a human-to-immunodeficient mouse testicular tissue transplantation [30]. In this study, the antibody stain for proacrosin was used to detect successful tubular implantation. Proacrosin is a marker of differentiated human spermatogonia (primary spermatocytes and spermatids) and would not have detected transplanted cells that had survived or propagated without further differentiation. On the other hand, Sofikitis et al. reported successful spermatogenesis from transplantation of human tissue into rat and mouse recipients [33]. In 2002, Nagano and colleagues reported on the use of anti-baboon testes antibody to identify the survival of human spermatogonia in mouse recipients for up to 6 months post-transplantation [32]. The same group then used human SSCs but found that these cells had not differentiated beyond spermatogonia 1 month after transplantation to the mouse testes. These interspecies transplantation studies seem to suggest, at present, that the vast phylogenetic distance between species (baboon to mouse and human to mouse) will likely translate into host testicular environment and donor spermatogonia incompatibilities that prohibit completion of spermatogenesis.

An alternative to germ cell transplantation is tissue grafting. Successful tissue xenografting of cryopreserved testicular specimen has been reported between various species [34–36] and non-human primates [37], Testicular tissue grafting entails placing a piece of the tissue, approximately 1 mm³ size, subcutaneously under the dorsal skin or the testicular bursa. However, the differentiation capacity of frozen spermatogonia showed xenografts of testicular tissues from immature rhesus monkeys as having a blockage in the spermatocyte stage [38]. Similar results were observed in frozen-thawed immature human testicular tissue grafted on the peritoneal bursa of mice testicles [39] where numerous pre-meiotic spermatocytes were observed; however, spermatid-like cells did not express meiotic or postmeiotic markers.

Severely Compromised Spermatogenesis

Spermatogeneic Arrest

Spermatogenic arrest in men occurs most frequently at the spermatogonium stage or during arrested meiosis, the latter being most common at the primary spermatocyte stage [40, 41]. Meiotic arrest, at the primary spermatocyte, is the most common in men with non-obstructive azoospermia [40].

Successful in vitro differentiation of spermatogenic cells into spermatids would be an extremely attractive option to treat some forms of male infertility caused by spermatogenic arrest [42, 43]. Various approaches to the challenge of in vitro spermatogenesis have been attempted, these mainly focusing on testicular cell cultures that are directed at achieving male germ cell differentiation. Among them, tissue culture, organ culture, and co-culture systems have been investigated extensively to assess which system can obtain meiotic or post-meiotic differentiation of these male germ cells [42–48]. Despite the report of a normal child being born after fertilization with germ cells from a man with maturation arrest at the primary spermatocyte stage [42], an effective procedure for the completion of spermatogenesis in vitro still remains elusive.

Treatment: In Vitro Spermatogenesis

In the early embryo, primordial germ cells (PGCs), the progenitors of gametes migrate from their yolk sac origin to nest within the somatic component of the gonad.

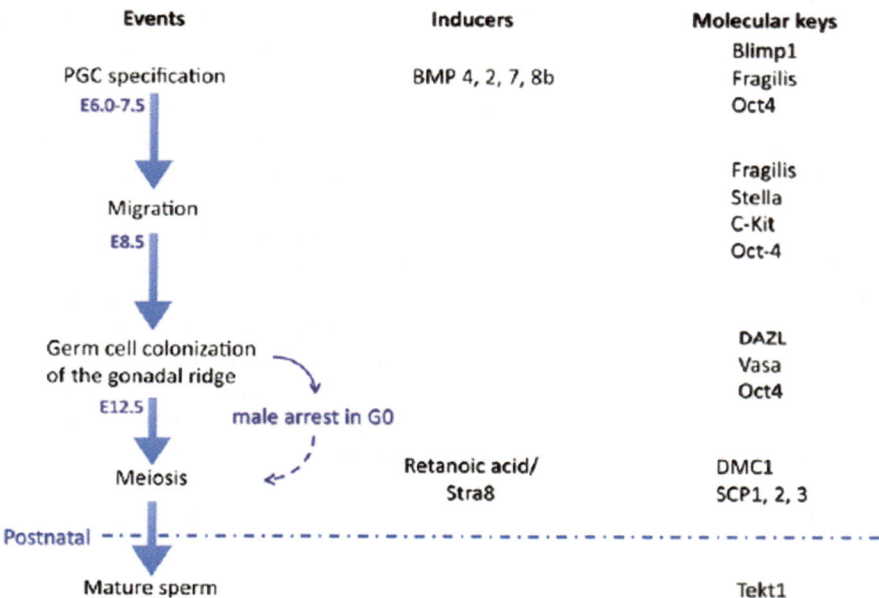

Fig. 8.2 The germline development: events and molecular keys. During in vivo development, BMP family stimulates a subset of epiblast cells that begin to express Fragilis and Blimp1 and become PGCs that migrate toward the gonadal ridge

They are one of the first discernable cell types in the mouse embryo. While their progress is essentially amebic in nature, several growth factors have been shown to influence the appearance, migration, survival, and proliferation. However, the signaling through which these cells localize in the genital ridges has not been identified.

Germ cells appear in mammals at the onset of gastrulation, soon after implantation of the embryo. In the mouse, PGCs are first distinguishable at the base of the allantois on embryonic day (E) 7.25 [49]. Lineage studies of epiblast cells show that mouse PGCs are specified by inductive interactions at the onset of gastrulation [50, 51]. Moreover, primary cultures of epiblast fragments from embryos at E5.5–E6.0 generate migrating PGCs when they are cocultured with extraembryonic ectoderm [52], and whole epiblasts from E6.0 embryos cultured on feeder cells expressing both bone morphogenic protein 4 (BMP4) and BMP8b will give rise to PGCs [53, 54]. These results reveal that bone morphogenic proteins (BMPs) derived from the extraembryonic ectoderm play crucial roles in PGC appearance in the proximal epiblast. PGCs will then migrate in the developing embryo toward the gonadal ridge, proliferate, and differentiate into gonocytes between E8.5–12.5 (Fig. 8.2).

To study the morphology, appearance, and migratory behavior of PGCs, in preliminary experiments we assessed embryo development as early as 6.5 dpc (Fig. 8.3a, b) with the embryo being 120×250 μm in size with an implantation site of 2×3 mm. After immunohistochemistry on 4–7 sections of 5 μm, a total of 14.6% (12/82) of the cells per embryo section expressed *fragilis* – a marker for premigratory PGCs. In a 12.5 dpc embryo (Fig. 8.4a) with a distinguishable gonadal ridge measuring

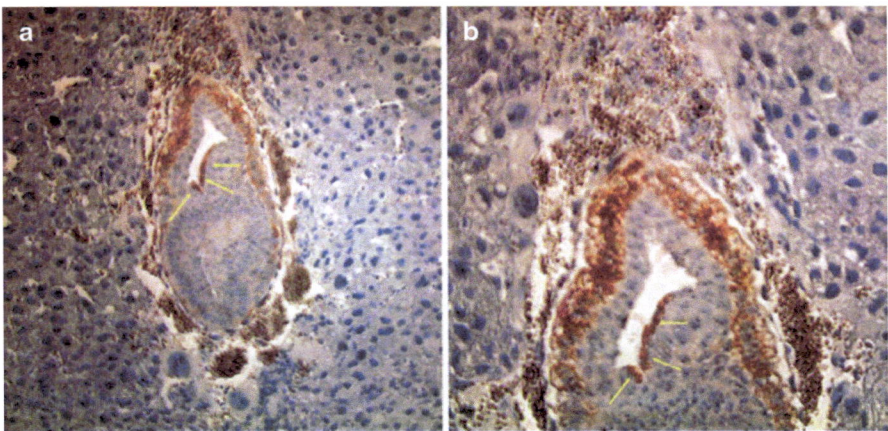

Fig. 8.3 Perfused embryo at 6.5 dpc. Primordial germ cells (PGCs) are labeled with VASA (*yellow arrows*) at (**a**) 100× and (**b**) 200×

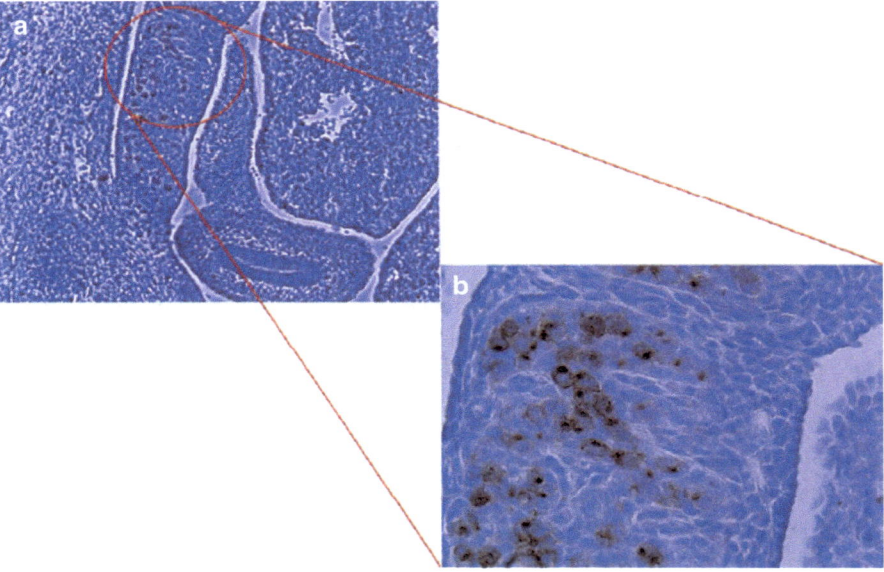

Fig. 8.4 Perfused embryo at 12.5 dpc. PGCs are labeled with VASA at (**a**) 100× and (**b**) 400×

25.0×87.5 μm, sections of this region evidenced 25.9% (45/174) VASA positive (Fig. 8.4b) post-migratory PGCs.

Gonocytes/SSCs self-renew and produce daughter cells that commit to differentiate into spermatozoa throughout adult life. SSCs can be identified unequivocally by a functional assay using a transplantation technique in which donor testis cells are injected into the seminiferous tubules of infertile recipient males [26, 55]. Under these conditions, only SSCs are able to generate foci of germ cells undergoing spermatogenesis [7].

One approach to the problem is cultivating SSCs under conditions that allow self-renewal and possibly induce differentiation. For this purpose, it is essential to establish a culture system with defined, experimentally modifiable characteristics. Glial cell line-derived neurotrophic factor (GDNF) is responsible for stimulating SSC renewal in rodents [56]. In addition, several BMP genes are expressed in the testis, in particular, BMP7 and BMP8 a/b transcripts [57, 58], suggesting that not only Sertoli cells, but also germ cells secrete molecules that can mediate paracrine interactions in the testis. In vitro co-culture experiments have shown that both BMP4 and BMP8B proteins function synergistically to induce PGC specification from the epiblast [54]. BMP2, another member of the BMP family, acts in cooperation with BMP4 as a third signal in PGC formation [53].

Induction of meiosis by retinoic acid (RA) released from the mesonephros helps determine the timing of meiosis according to the gender. In males, induction is delayed until the post-natal period and RA has been postulated as being secreted by Sertoli cells [59]. RA plays a role in the regulation of meiosis and is required to be present to have the *Stra8* (stimulated by RA 8) gene to become expressed. The Stra8 encodes a cytoplasmic protein and is expressed specific to the developing male gonad during mouse embryogenesis and in the adult is restricted to the premeiotic germ cells.

Another important key to the full establishment of SSC in vitro is the incorporation of an appropriate feeder layer [60]. In the 1960s, Steinberger and associates found in numerous experimental trials that culturing germ cells that were mechanically or enzymatically dissociated from the testis had their survival directly dependent upon sertoli cell co-culture [61–63]. Recently, differentiation of rat germ cells into spermatids from preleptotene spermatocytes has been documented using Sertoli cells as a feeder layer [64].

As another variant, we attempted to isolate and select SSCs by utilizing magnetic activated cell sorting (Fig. 8.5). In order to obtain the maximal yield of SSCs, testes were retrieved from both neonate and adult B_6D_2-F_1 mice. The testicular cell suspension were layered onto a 30% density gradient and centrifuged for 8 min. CD90 (Thy 1.2) antibody was elected as a marker to identify SSC fractions. The excess antibody was removed and the resulting cell suspension was loaded onto a separation column within a magnetic field. Fourteen testes retrieved from 7-6 dpc mice had an average weight of 5.0 mg±0.001. Sorting of six dissected testes (795,000 testicular cells) by magnetic separation yielded ~43,500 CD90 positive cells with a >80% enrichment rate ($P<0.0001$). After seeding the germ cells on a feeder layer (Fig. 8.6), one putative SSC colony, confirmed by alkaline phosphatase (AP) activity/VASA positivity, proliferated up to day 9 of culture (Fig. 8.7).

The remaining eight testes provided about 1.1×10^6 cells with ~3% germ cells retrieved after enzymatic digestion. The unselected suspensions of testicular cells consisting of a mixture of SSCs, sertoli cells, and interstitial cells were plated under three culture conditions: (1) GDNF+bFGF or (2) FSH+bFGF or (3) GDNF+bFGF+LIF. The following day, epithelial cells had become attached to the dish to form a confluent monolayer with scattered adherent germ cells (Fig. 8.8). On

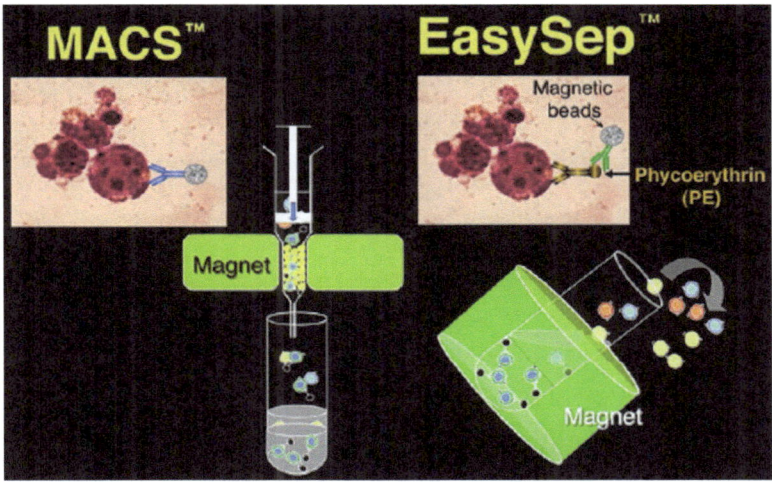

Fig. 8.5 Magnetic cell sorting by MACS™ or EasySep™. Cell suspensions were first processed by density gradient centrifugation and then sorted by incubation with monoclonal Thy-1 (CD90) magnetic-bead conjugated antibody

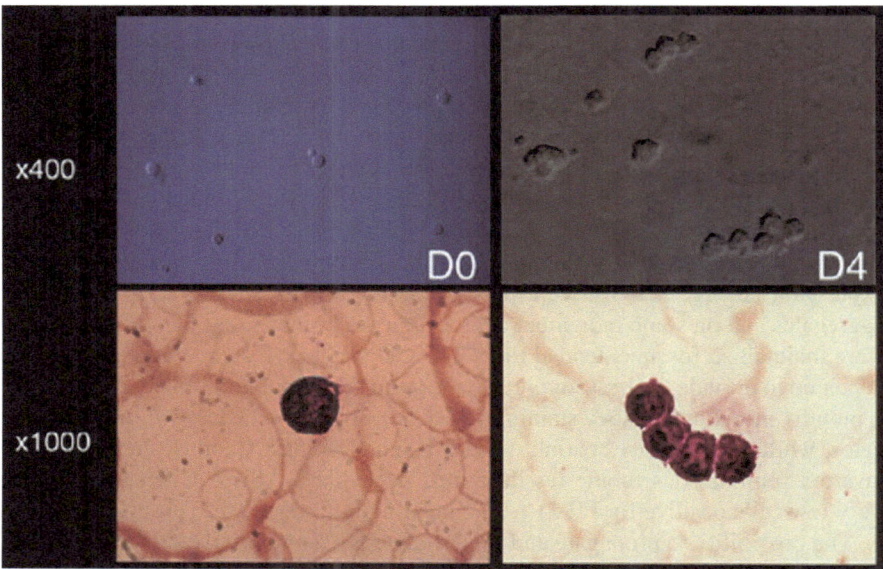

Fig. 8.6 After germ cells enrichment, cells were seeded (D0 of culture) while others were stained for morphological analysis. At D4 of culture, some germ cell proliferation was observed

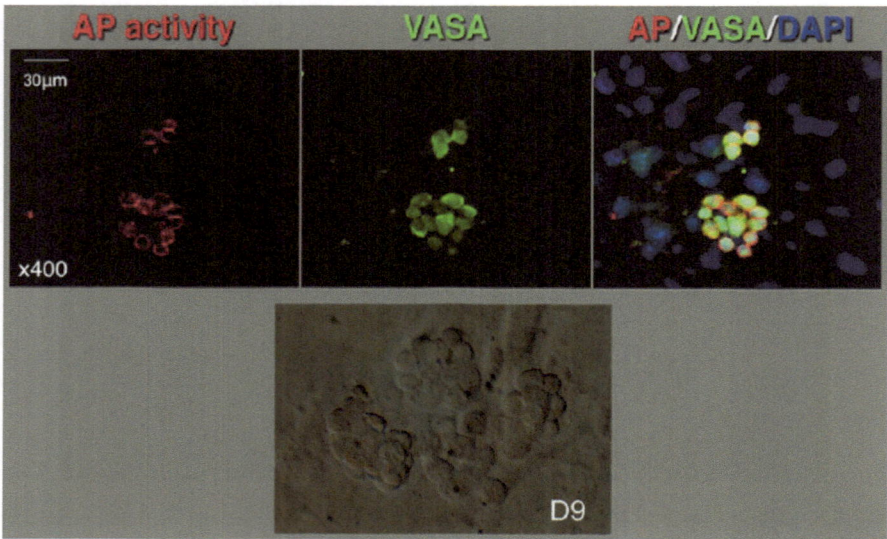

Fig. 8.7 Double expression of pluripotent markers alkaline phosphatase activity and VASA were assessed on plating D4 of SSC colonies. Nuclei were counterstained with DAPI. Putative colonies proliferated up to day 9 of culture

D3, cell clumps of ~25 μm diameter were uniformly distributed in the dish. Over the subsequent days, there was a spontaneous and progressive migration of the somatic cells around each germ cell, creating dispersed cell clumps. In every media combination, the size of the cell aggregates gradually increased to reach a diameter of ~100 μm by D10 (Fig. 8.9). In GDNF + bFGF and GDNF + bFGF + LIF, the aggregates remained as tightly packed entities, whereas the clumps in FSH + bFGF had started to spread. When LIF was removed, cell aggregates behaved similarly to those in FSH + bFGF. The putative germ cells in these tridimensional seminiferous tubule-like structures were then monitored for up to 75 days. Such cellular aggregates had a steady linear growth up to D56 and then plateaued thereafter (Fig. 8.10). Aggregates were fixed at different times to assess presence of germ cells. VASA expression and AP activity were present mainly on the outside periphery of the aggregates and on some individual cells within them, indicating that the germ cells were maintained for an extended time. The GDNF + bFGF supplemented with LIF appeared to provide the best support for the germ cell population. Interestingly, after 2 months in culture, VASA staining revealed only a few PGCs within the aggregates, while the majority coating them, presumably SSCs, adopted the form of an inverted seminiferous tubule (as described by Schlatt, ASRM Course 21, 2007). Very few cells positive for FE-J1 (~1%; post-meiotic stage) were observed [65].

The possibility to propagate and immortalize SSCs in vitro would allow the creation of colonies for a specific individual suffering from germ cell maturation arrest, and more importantly, patients suffering from childhood cancer. These cells can be maintained, cryopreserved, and made available for transplantation to the testes of individuals ready to start a family.

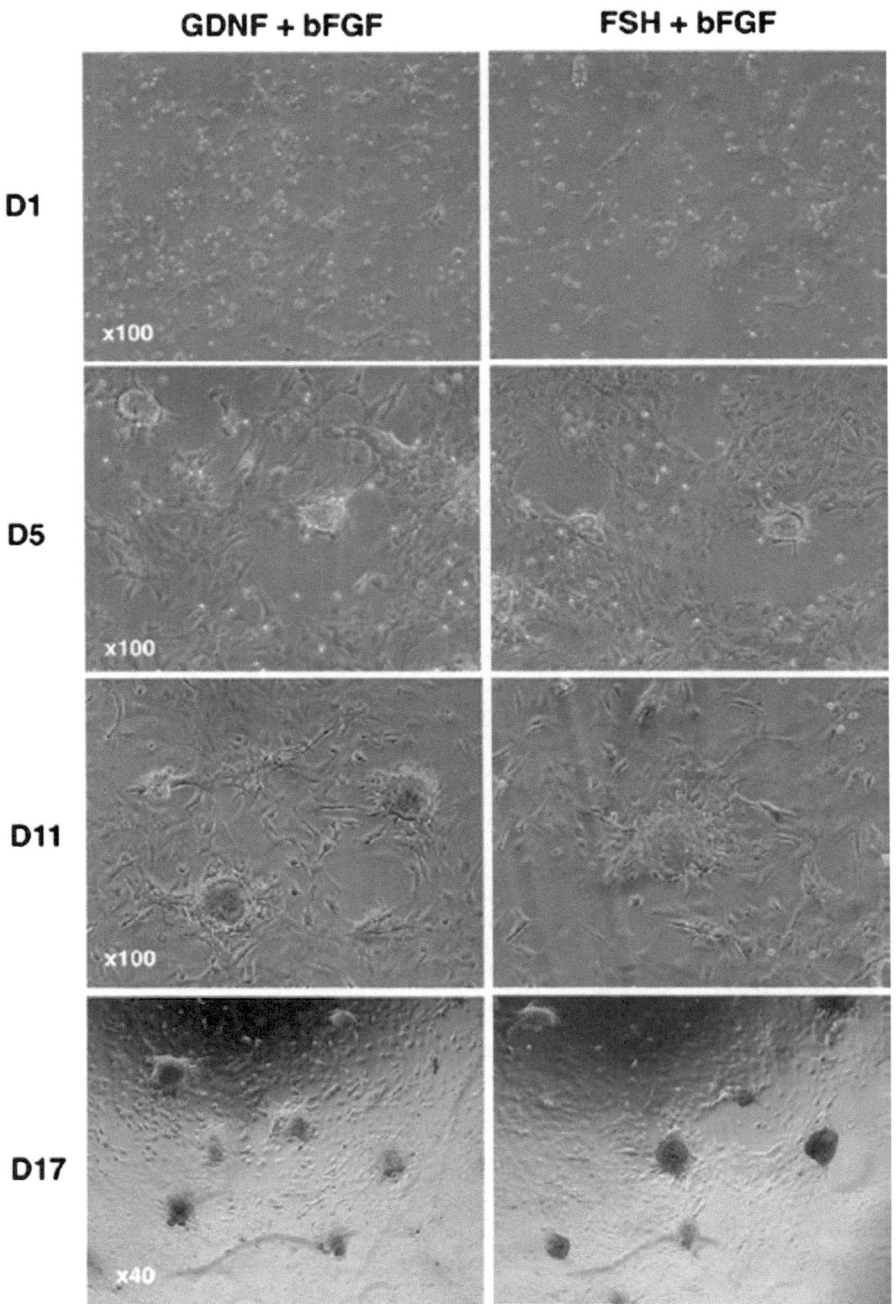

Fig. 8.8 Neonatal testicular cells started to form cell clumps on D1 of culture in a serum-free medium supplemented with growth factors. Cells began to adhere to the dish and form scattered clusters (D5). These cell aggregates progressively grew over time as seen on frames D11 and D17

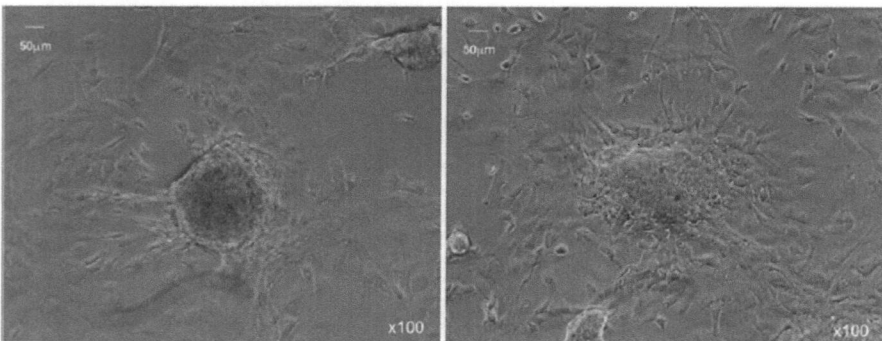

Fig. 8.9 Extended culture of unselected testicular cells in serum-free medium supplemented different growth factors. On D17, in GDNF+bFGF were tightly packed while FSH+bFGF began to spread over the dish

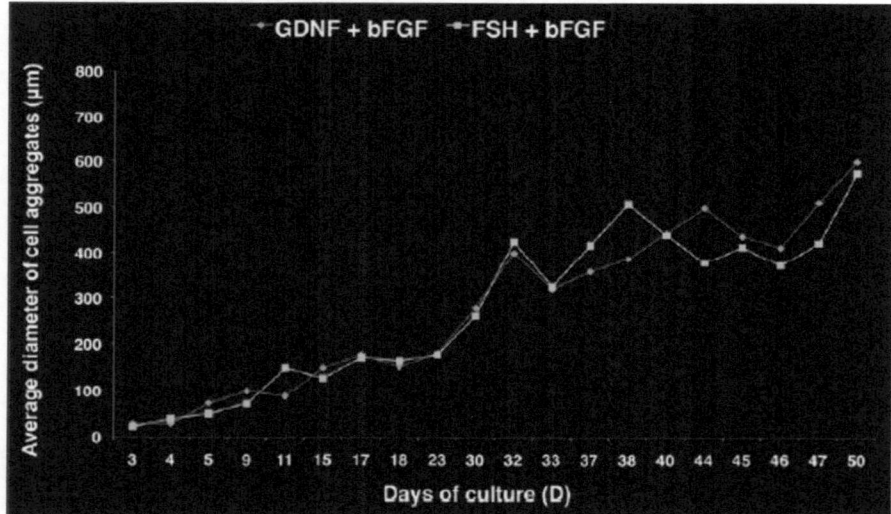

Fig. 8.10 Size of the cell aggregates progressively increased with time

Germ Cell Aplasia

SCO syndrome, also called germ cell aplasia, describes a condition of the testes in which germ cells are absent and only Sertoli cells populate the seminiferous tubules. SCO syndrome presents during the evaluation of azoospermia in couples having difficulty in initiating a pregnancy.

These men typically present with infertility as the only symptom, and most present between age 20 and 40 years, and are found to have no sperm in the ejaculate. The male urologic physical examinations are often unremarkable, and the diagnosis is made from a testicular biopsy. While investigation of the basis for the SCO syndrome is ongoing, the etiology and mechanism of this process are currently unknown. Some examples of diseases and disorders characterized by SCO and Leydig cell hyperplasia are Klinefelter's Syndrome, mumps orchitis, deletion of the Y chromosome at the AZFa or AZFb regions, and men/children that had anti-neoplastic therapy with radiation or chemotherapy.

The prevalence of the SCO syndrome in the overall population is extremely low. Of the 10% of US couples affected by infertility, approximately 30% have a pure male factor as the underlying cause, and another 20% have a combined male and female factor. Although precise figures are difficult to obtain, less than 5–10% of these infertile men have SCO syndrome. SCO syndrome has no known racial predilection; however, SCO is more common in white men. There are no known effective treatment(s) at the present time.

Future Options

During embryonic stem cell (ESC) differentiation through embryoid body (EB) development in vitro, tissue-specific genes and proteins are differentially expressed in patterns that in some way mimic mouse embryogenesis in vivo [66]. ESCs have shown the ability to recapitulate features of embryonic development by spontaneously forming somatic lineages in culture. While ESCs have been maintained in monolayer cultures, sporadic oocyte-like structures have been identified that presumably develop into a structure resembling a blastocyst [67]. Follicle-like structures were also obtained by feeding EBs with conditioned medium isolated from cultures of mouse neonatal testicular tissue [68]. These structures were not, however, capable of developing further.

Growth factors such as leukemia inhibiting factor (LIF) and the BMP derivatives [69] are considered essential to grant their survival to get germ cell proliferation and self-renewal, while other factors such as RA would help them to enter meiosis [70–73]. By changing the concentration of BMPs (BMP4, 7, and 8b; 1 or 10 μM RA) and using EB formation, we attempted to recreate a seminiferous tubule-like environment [74]. We observed that 10 μm RA had the highest proportion of germ cell appearance on day 3 and was able to maintain up to day 20. In a Brazilian study, using commercially made neurobasal medium supplemented with B27 were presumably able to obtain sperm in 10 days and oocytes in 25 days of culture in the same EB [75].

The inherent characteristics of ESCs – their capacity to differentiate into cells of all three germ layers and grow indefinitely [76–78], presented an enormous potential for the field of regenerative medicine. A reserve of patient-specific pluripotent cell cultures, with the potential to differentiate into germline cells, remains an idealized goal in the minds of ART researchers and clinicians. Though the possibilities for ESC-based therapy could be enmeshed in ethical controversy, alternative treatment options have shown promise in recent years. Specifically, overexpression of the transcription factors Oct4, Klf4, Sox2, and c-Myc can induce somatic cells to exhibit molecular and functional similarities to ESCs, aptly designated as induced pluripotent stem cells (iPSCs) [78]. Initial efforts promoted the introduction of candidate genes into mouse embryonic fibroblasts (MEFs) via retroviral transduction, with primary culture on STO feeder cells in ES cell medium [78]. However, further efforts have revealed a need for alternative sources, as the reprogramming of fibroblasts proves to be a relatively slow (2–3 weeks) and inefficient (<1%) process [79].

Derivation of iPSCs from human peripheral blood cells, specifically T lymphocytes and myeloid cells, is one such alternative showing recent promise. Utilizing a polycistronic vector encoding the four standard transcription factors, followed by transfer onto MEF feeder layers and ensuing culture incubation, blood cells can give rise to expandable and stable iPSC colonies [80]. These colonies stain for pluripotency markers such as Oct4, Nanog, and Tral-81, and are capable of in vitro differentiation into cells of the three germ layers [81]. Exploring alternative candidate sources such as blood for pluripotent induction provides much of the contemporary focus in regenerative medicine [81–83]. Isolating cells that exhibit higher proliferation rates, better long-term growth potentials and convenient accessibility remains an important step to making iPSC technology more broadly applicable.

Conclusions

Aimed At: Azoospermic Men

ICSI is the most effective treatment currently employed for severe male factor infertility but this requires the presence of fully developed spermatozoa. However, for men suffering from complete spermatogenic arrest treatment is still elusive. In mammals, millions of spermatozoa are produced daily, ultimately from spermatogonia stem cells. These precursors of spermatogonia depend not only on specific growth factors for survival and proliferation, but also on the close relationship with the enveloping presumptive sertoli cells which seems to prevent apoptosis of these germ cells. The ability to propagate and immortalize SSCs in vitro would allow the creation of colonies, which on intra-testicular transplantation might be able to repopulate the germinal epithelium of azoospermic men with SCO syndrome.

Men with germ cell aplasia as well as men and pre-pubescent children surviving cancer who had undergone chemotherapy treatment require the transplantation of germ cells to re-initiate the spermatogenic process. Since chemotherapy involves the use of extremely cytotoxic chemicals, it can cause tremendous harm to germ cells, often rendering the patient sterile.

In this regard, significant strides have been accomplished in experimental animals through spermatogonia transplantation that offers an opportunity for future clinical application in the human. A handful of researchers have been able to recolonize the basal membrane of seminiferous tubules by injecting SSCs, therefore reestablishing spermatogenesis in autologous transfers. The so derived viable spermatozoa were capable of fertilizing mature oocytes resulting in the production of progeny [84]. As experiments involving human subjects are limited by ethical constraints, xenogeneic transplantation using experimental models has helped provide an adequate environment to study the different stages of germ cell differentiation. Attempts have been made to transplant human germ cell suspensions into the testes of immunodeficient mice, but unfortunately without success in the case of fresh cell suspensions [31] and with limited success in the case of cryopreserved cells [33]. Xenografting of human testicular tissue pieces has also met with limited success [39, 85, 86]. Both xenotransplantation and -grafting are blocked at premeiotic stages and have no differentiation capacity to the later spermiogeneic stages.

What Can Be Done in the Future

At the present time, cryopreservation methods are limited to utilizing standard freezing methods either with DMSO or glycerol [87–89]. Another method that has been attempted to assess if spermatogonial cells have better survival rates is vitrification [90]. However, the main limitation of cryopreservation is the possible damage to the spermatogonia thereby limiting its survival as well as its ability to self renew. In this instance, the addition of growth factors may help germ cell survival, proliferation, and eventual differentiation to mature gametes.

Where Future Research Is Heading

With the availability of embryonic stem cells the possibility to recreate a male germ cell niche, through the tridimensional model represented by the embryoid bodies, appears to be concrete. While these studies indicate that alternative sources of gametes are not merely the result of science fiction but a reality, further advances on the epigenesis of the neo-gametes are welcome.

Moreover, the possibility to propagate a male genome would provide an alternative means through which to consistently obtain conceptuses. We attempted to replicate a haploid male genome in the mouse [75, 91] and generate conceptuses capable of undergoing full-term development, so as to enhance the reproductive performance of a single spermatozoon.

When all the techniques are refined, scientists and physicians can develop a possible treatment that will restore fecundity in a sterile man and reproductive medicine will forever be changed. Thus, SSCs and their subsequent culturing in vitro is the future of reproductive medicine, finally giving hope to infertile couples who wish for nothing more than starting a family of their own.

Acknowledgments We are grateful to the clinicians, scientists, embryologists, and nursing staff of The Ronald O. Perelman and Claudia Cohen Center for Reproductive Medicine. We are thankful to Dr. J. Michael Bedford for his critical review, Dr. Noriko Tanaka for establishing the magnetic cell sorting, and for Justin Kocent for literature search on iPS. Queenie V. Neri was funded by the grant ULI RR024996 of the Clinical and Translational Science Center at Weill Cornell Medical College.

References

1. Stefankiewicz J, Kurzawa R, Drozdzik M. Environmental factors disturbing fertility of men. Ginekol Pol. 2006;77:163–9.
2. Wright VC, Chang J, Jeng G, Macaluso M. Assisted reproductive technology surveillance – United States, 2003. MMWR Surveill Summ. 2006;55:1–22.
3. Chan PT, Schlegel PN. Nonobstructive azoospermia. Curr Opin Urol. 2000;10:617–24.
4. Palermo G, Joris H, Devroey P, Van Steirteghem AC. Pregnancies after intracytoplasmic injection of single spermatozoon into an oocyte. Lancet. 1992;340:17–8.
5. Palermo GD, Schlegel PN, Colombero LT, Zaninovic N, Moy F, Rosenwaks Z. Aggressive sperm immobilization prior to intracytoplasmic sperm injection with immature spermatozoa improves fertilization and pregnancy rates. Hum Reprod. 1996;11:1023–9.
6. Tanaka H, Baba T. Gene expression in spermiogenesis. Cell Mol Life Sci. 2005;62:344–54.
7. Brinster RL. Germline stem cell transplantation and transgenesis. Science. 2002;296:2174–6.
8. Dym M, He Z, Jiang J, Pant D, Kokkinaki M. Spermatogonial stem cells: unlimited potential. Reprod Fertil Dev. 2009;21:15–21.
9. Bleyer A, Barr R. Cancer in young adults 20 to 39 years of age: overview. Semin Oncol. 2009;36:194–206.
10. Schover LR, Rybicki LA, Martin BA, Bringelsen KA. Having children after cancer. A pilot survey of survivors' attitudes and experiences. Cancer. 1999;86:697–709.
11. Tournaye H, Goossens E, Verheyen G, Frederickx V, De Block G, Devroey P, et al. Preserving the reproductive potential of men and boys with cancer: current concepts and future prospects. Hum Reprod Update. 2004;10:525–32.
12. Schrader M, Heicappell R, Muller M, Straub B, Miller K. Impact of chemotherapy on male fertility. Onkologie. 2001;24:326–30.
13. Zapzalka DM, Redmon JB, Pryor JL. A survey of oncologists regarding sperm cryopreservation and assisted reproductive techniques for male cancer patients. Cancer. 1999;86:1812–7.
14. Allen C, Keane D, Harrison RF. A survey of Irish consultants regarding awareness of sperm freezing and assisted reproduction. Ir Med J. 2003;96:23–5.
15. Schrader M, Muller M, Straub B, Miller K. Testicular sperm extraction in azoospermic patients with gonadal germ cell tumors prior to chemotherapy – a new therapy option. Asian J Androl. 2002;4:9–15.
16. Carson SA, Gentry WL, Smith AL, Buster JE. Feasibility of semen collection and cryopreservation during chemotherapy. Hum Reprod. 1991;6:992–4.
17. Robbins WA, Meistrich ML, Moore D, Hagemeister FB, Weier HU, Cassel MJ, et al. Chemotherapy induces transient sex chromosomal and autosomal aneuploidy in human sperm. Nat Genet. 1997;16:74–8.
18. Brandriff BF, Meistrich ML, Gordon LA, Carrano AV, Liang JC. Chromosomal damage in sperm of patients surviving Hodgkin's disease following MOPP (nitrogen mustard, vincristine, procarbazine, and prednisone) therapy with and without radiotherapy. Hum Genet. 1994;93:295–9.
19. Foresta C, Bettella A, Marin P, Galeazzi C, Merico M, Scandellari C. Analysis of sperm aneuploidy in infertile subjects after chemotherapy treatment. Ann Ital Med Int. 2000;15:189–94.
20. Berthelsen JG, Skakkebaek NE. Gonadal function in men with testis cancer. Fertil Steril. 1983;39:68–75.
21. Baniel J, Sella A. Sperm extraction at orchiectomy for testis cancer. Fertil Steril. 2001;75:260–2.

22. Hourvitz A, Goldschlag DE, Davis OK, Gosden LV, Palermo GD, Rosenwaks Z. Intracytoplasmic sperm injection (ICSI) using cryopreserved sperm from men with malignant neoplasm yields high pregnancy rates. Fertil Steril. 2008;90:557–63.
23. Hawkins MM, Stevens MC. The long-term survivors. Br Med Bull. 1996;52:898–923.
24. Bleyer WA. The impact of childhood cancer on the United States and the world. CA Cancer J Clin. 1990;40:355–67.
25. Brinster RL, Avarbock MR. Germline transmission of donor haplotype following spermatogonial transplantation. Proc Natl Acad Sci USA. 1994;91:11303–7.
26. Brinster RL, Nagano M. Spermatogonial stem cell transplantation, cryopreservation and culture. Semin Cell Dev Biol. 1998;9:401–9.
27. Clouthier DE, Avarbock MR, Maika SD, Hammer RE, Brinster RL. Rat spermatogenesis in mouse testis. Nature. 1996;381:418–21.
28. Ogawa T, Dobrinski I, Avarbock MR, Brinster RL. Xenogeneic spermatogenesis following transplantation of hamster germ cells to mouse testes. Biol Reprod. 1999;60:515–21.
29. Dobrinski I, Avarbock MR, Brinster RL. Transplantation of germ cells from rabbits and dogs into mouse testes. Biol Reprod. 1999;61:1331–9.
30. Reis MM, Tsai MC, Schlegel PN, Feliciano M, RaffaeHi R, Rosenwaks Z, et al. Xenogeneic transplantation of human spermatogonia. Zygote. 2000;8:97–105.
31. Nagano M, Brinster CJ, Orwig KE, Ryu BY, Avarbock MR, Brinster RL. Transgenic mice produced by retroviral transduction of male germ-line stem cells. Proc Natl Acad Sci USA. 2001;98:13090–5.
32. Nagano M, Watson DJ, Ryu BY, Wolfe JH, Brinster RL. Lentiviral vector transduction of male germ line stem cells in mice. FEBS Lett. 2002;524:111–5.
33. Sofikitis N, Ono K, Yamamoto Y, Papadopoulos H, Miyagawa I. Influence of the male reproductive tract on the reproductive potential of round spermatids abnormally released from the seminiferous epithelium. Hum Reprod. 1999;14:1998–2006.
34. Honaramooz A, Snedaker A, Boiani M, Scholer H, Dobrinski I, Schlatt S. Sperm from neonatal mammalian testes grafted in mice. Nature. 2002;418:778–81.
35. Schlatt S, Kim SS, Gosden R. Spermatogenesis and steroidogenesis in mouse, hamster and monkey testicular tissue after cryopreservation and heterotopic grafting to castrated hosts. Reproduction. 2002;124:339–46.
36. Shinohara T, Inoue K, Ogonuki N, Kanatsu-Shinohara M, Miki H, Nakata K, et al. Birth of offspring following transplantation of cryopreserved immature testicular pieces and in-vitro microinsemination. Hum Reprod. 2002;17:3039–45.
37. Orwig KE, Schlatt S. Cryopreservation and transplantation of spermatogonia and testicular tissue for preservation of male fertility. J Natl Cancer Inst Monogr. 2005;34:51–6.
38. Jahnukainen K, Ehmcke J, Hergenrother SD, Schlatt S. Effect of cold storage and cryopreservation of immature non-human primate testicular tissue on spermatogonial stem cell potential in xenografts. Hum Reprod. 2007;22:1060–7.
39. Wyns C, Van Langendonckt A, Wese FX, Donnez J, Curaba M. Long-term spermatogonial survival in cryopreserved and xenografted immature human testicular tissue. Hum Reprod. 2008;23:2402–14.
40. Martin-du Pan RC, Campana A. Physiopathology of spermatogenic arrest. Fertil Steril. 1993;60:937–46.
41. Sasagawal YH, Suzuki Y, Tateno T, Ichiyanagi O, Kobayashi T, et al. Reevaluation of testicular biopsies of males with nonobstructive azoospermia in assisted reproductive technology. Arch Androl. 2001;46:79–83.
42. Tesarik J, Bahceci M, Ozcan C, Greco E, Mendoza C. Restoration of fertility by in-vitro spermatogenesis. Lancet. 1999;353:555–6.
43. Staub C. A century of research on mammalian male germ cell meiotic differentiation in vitro. J Androl. 2001;22:911–26.
44. Hue D, Staub C, Perrard-Sapori MH, Weiss M, Nicolle JC, Vigier M, et al. Meiotic differentiation of germinal cells in three-week cultures of whole cell population from rat seminiferous tubules. Biol Reprod. 1998;59:379–87.

45. Cremades N, Bernabeu R, Barros A, Sousa M. In-vitro maturation of round spermatids using co-culture on Vero cells. Hum Reprod. 1999;14:1287–93.
46. Sousa M, Cremades N, Alves C, Silva J, Barros A. Developmental potential of human spermatogenic cells co-cultured with Sertoli cells. Hum Reprod. 2002;17:161–72.
47. Tesarik J, Nagy P, Abdelmassih R, Greco E, Mendoza C. Pharmacological concentrations of follicle-stimulating hormone and testosterone improve the efficacy of in vitro germ cell differentiation in men with maturation arrest. Fertil Steril. 2002;77:245–51.
48. Tanaka A, Nagayoshi M, Awata S, Mawatari Y, Tanaka I, Kusunoki H. Completion of meiosis in human primary spermatocytes through in vitro coculture with Vero cells. Fertil Steril. 2003;79(Suppl 1):795–801.
49. Ginsburg M, Snow MH, McLaren A. Primordial germ cells in the mouse embryo during gastrulation. Development. 1990;110:521–8.
50. Lawson KA, Hage WJ. Clonal analysis of the origin of primordial germ cells in the mouse. Ciba Found Symp. 1994;182:68–84. Discussion 84–91.
51. Saitou M, Barton SC, Surani MA. A molecular programme for the specification of germ cell fate in mice. Nature. 2002;418:293–300.
52. Yoshimizu T, Obinata M, Matsui Y. Stage-specific tissue and cell interactions play key roles in mouse germ cell specification. Development. 2001;128:481–90.
53. Ying Y, Zhao GQ. Cooperation of endoderm-derived BMP2 and extraembryonic ectoderm-derived BMP4 in primordial germ cell generation in the mouse. Dev Biol. 2001;232:484–92.
54. Ying Y, Qi X, Zhao GQ. Induction of primordial germ cells from murine epiblasts by synergistic action of BMP4 and BMP8B signaling pathways. Proc Natl Acad Sci USA. 2001;98:7858–62.
55. Brinster RL, Zimmermann JW. Spermatogenesis following male germ-cell transplantation. Proc Natl Acad Sci USA. 1994;91:11298–302.
56. Schmidt JA, Avarbock MR, Tobias JW, Brinster RL. Identification of glial cell line-derived neurotrophic factor-regulated genes important for spermatogonial stem cell self-renewal in the rat. Biol Reprod. 2009;81:56–66.
57. Zhao GQ, Deng K, Labosky PA, Liaw L, Hogan BL. The gene encoding bone morphogenetic protein 8B is required for the initiation and maintenance of spermatogenesis in the mouse. Genes Dev. 1996;10:1657–69.
58. Zhao GQ, Chen YX, Liu XM, Xu Z, Qi X. Mutation in Bmp7 exacerbates the phenotype of Bmp8a mutants in spermatogenesis and epididymis. Dev Biol. 2001;240:212–22.
59. Bowles J, Knight D, Smith C, Wilhelm D, Richman J, Mamiya S, et al. Retinoid signaling determines germ cell fate in mice. Science. 2006;312:596–600.
60. Kanatsu-Shinohara M, Ogonuki N, Inoue K, Miki H, Ogura A, Toyokuni S, et al. Long-term proliferation in culture and germline transmission of mouse male germline stem cells. Biol Reprod. 2003;69:612–6.
61. Steinberger A, Steinberger E, Perloff WH. Mammalian testes in organ culture. Exp Cell Res. 1964;36:19–27.
62. Steinberger E, Steinberger A, Perloff WH. Studies on growth in organ culture of testicular tissue from rats of various ages. Anat Rec. 1964;148:581–9.
63. Rassoulzadegan M, Paquis-Flucklinger V, Bertino B, Sage J, Jasin M, Miyagawa K, et al. Transmeiotic differentiation of male germ cells in culture. Cell. 1993;75:997–1006.
64. Ogawa T, Kita K, Kubota Y. Proliferation of spermatogonial stem cells and spermatogenesis in vitro. Reprod Med Biol. 2006;5:169–74.
65. Neri QV, Tanaka N, Takeuchi T, Toschi M, Rosenwaks Z, Palermo GD. Propagation and maturation of male gonocytes in vitro. Fertil Steril. 2006;86:sl4.
66. Rohwedel J, Guan K, Wobus AM. Induction of cellular differentiation by retinoic acid in vitro. Cells Tissues Organs. 1999;165:190–202.
67. Hubner K, Fuhrmann G, Christenson LK, Kehler J, Reinbold R, De La Fuente R, et al. Derivation of oocytes from mouse embryonic stem cells. Science. 2003;300:1251–6.
68. Lacham-Kaplan O, Chy H, Trounson A. Testicular cell conditioned medium supports differentiation of embryonic stem cells into ovarian structures containing oocytes. Stem Cells. 2006;24:266–73.

69. Clark AT, Bodnar MS, Fox M, Rodriquez RT, Abeyta MJ, Firpo MT, et al. Spontaneous differentiation of germ cells from human embryonic stem cells in vitro. Hum Mol Genet. 2004;13:727–39.
70. Geijsen N, Horoschak M, Kim K, Gribnau J, Eggan K, Daley GQ. Derivation of embryonic germ cells and male gametes from embryonic stem cells. Nature. 2004;427:148–54.
71. Guan K, Nayernia K, Maier LS, Wagner S, Dressel R, Lee JH, et al. Pluripotency of spermatogonial stem cells from adult mouse testis. Nature. 2006;440:1199–203.
72. Nayernia K, Nolte J, Michelmann HW, Lee JH, Rathsack K, Drusenheimer N, et al. In vitro-differentiated embryonic stem cells give rise to male gametes that can generate offspring mice. Dev Cell. 2006;11:125–32.
73. Izadyar F, Pau F, Marh J, Slepko N, Wang T, Gonzalez R, et al. Generation of multipotent cell lines from a distinct population of male germ line stem cells. Reproduction. 2008;135:771–84.
74. Palermo GD, Neri QV, Takeuchi T, Rosenwaks Z. ICSI: where we have been and where we are going. Semin Reprod Med. 2009;27:191–201.
75. Kerkis A, Fonseca SA, Serafim RC, LavagnoHi TM, Abdelmassih S, Abdelmassih R, et al. In vitro differentiation of male mouse embryonic stem cells into both presumptive sperm cells and oocytes. Cloning Stem Cells. 2007;9:535–48.
76. Evans MJ, Kaufman MH. Establishment in culture of pluripotential cells from mouse embryos. Nature. 1981;292:154–6.
77. Martin GR. Isolation of a pluripotent cell line from early mouse embryos cultured in medium conditioned by teratocarcinoma stem cells. Proc Natl Acad Sci USA. 1981;78:7634–8.
78. Thomson JA, Itskovitz-Eldor J, Shapiro SS, Waknitz MA, Swiergiel JJ, Marshall VS, et al. Embryonic stem cell lines derived from human blastocysts. Science. 1998;282:1145–7.
79. Takahashi K, Yamanaka S. Induction of pluripotent stem cells from mouse embryonic and adult fibroblast cultures by defined factors. Cell. 2006;126:663–76.
80. Hochedlinger K, Plath K. Epigenetic reprogramming and induced pluripotency. Development. 2009;136:509–23.
81. Staerk J, Dawlaty MM, Gao Q, Maetzel D, Hanna J, Sommer CA, et al. Reprogramming of human peripheral blood cells to induced pluripotent stem cells. Cell Stem Cell. 2010;7:20–4.
82. Loh YH, Hartung O, Li H, Guo C, Sahalie JM, Manos PD, et al. Reprogramming of T cells from human peripheral blood. Cell Stem Cell. 2010;7:15–9.
83. Seki T, Yuasa S, Oda M, Egashira T, Yae K, Kusumoto D, et al. Generation of induced pluripotent stem cells from human terminally differentiated circulating T cells. Cell Stem Cell. 2010;7:11–4.
84. Kubota H, Avarbock MR, Schmidt JA, Brinster RL. Spermatogonial stem cells derived from infertile Wv/Wv mice self-renew in vitro and generate progeny following transplantation. Biol Reprod. 2009;81:293–301.
85. Geens M, De Block G, Goossens E, Frederickx V, Van Steirteghem A, Tournaye H. Spermatogonial survival after grafting human testicular tissue to immunodeficient mice. Hum Reprod. 2006;21:390–6.
86. Schlatt S, Honaramooz A, Ehmcke J, Goebell PJ, Rubben H, Dhir R, et al. Limited survival of adult human testicular tissue as ectopic xenograft. Hum Reprod. 2006;21:384–9.
87. Kvist K, Thorup J, Byskov AG, Hoyer PE, Mollgard K, Yding Andersen C. Cryopreservation of intact testicular tissue from boys with cryptorchidism. Hum Reprod. 2006;21:484–91.
88. Keros V, Hultenby K, Borgstrom B, Fridstrom M, Jahnukainen K, Hovatta O. Methods of cryopreservation of testicular tissue with viable spermatogonia in pre-pubertal boys undergoing gonadotoxic cancer treatment. Hum Reprod. 2007;22:1384–95.
89. Wyns C, Curaba M, Martinez-Madrid B, Van Langendonckt A, Francois-Xavier W, Donnez J. Spermatogonial survival after cryopreservation and short-term orthotopic immature human cryptorchid testicular tissue grafting to immunodeficient mice. Hum Reprod. 2007;22:1603–11.
90. Curaba M, Verleysen M, Amorim CA, Dolmans MM, Van Langendonckt A, Hovatta O, et al. Cryopreservation of prepubertal mouse testicular tissue by vitrification. Fertil Steril. 2011;95(4):1229–34.
91. Takeuchi T, Neri QV, Palermo GD. Male gamete empowerment. Ann N Y Acad Sci. 2008;1127:64–6.

Chapter 9
Testicular Tissue Transplantation for Fertility Preservation

Jose R. Rodriguez-Sosa, Stefan Schlatt, and Ina Dobrinski

Male fertility is based on the efficient production of sperm during the adult life of a male. Sperm are generated through a complex process of cell proliferation and differentiation known as spermatogenesis. In this process, spermatogonia undergo successive divisions to give rise to spermatocytes, in which genetic material is recombined and segregated through meiosis. Resulting haploid spermatids initiate a series of morphological changes to finally transform into sperm structurally equipped for the efficient delivery of the genetic material at fertilization [1]. Spermatogenesis occurs as a continuous process, assuring virtually unlimited sperm production during adult life. This process is sustained by spermatogonial stem cells (SSCs), which not only give rise to differentiating spermatogonia during the proliferative phase of spermatogenesis, but also self-renew to maintain the reserve stem cell pool [2]. Whether SSCs undergo asymmetric division resulting in stem cells and differentiated daughter cells is still a subject of investigation [3]. Recent evidence also points to the possibility that the decision to remain a stem cell or to enter the differentiation pathway can occur through fragmentation of clones of spermatogonia that will enter different pathways. This phenomenon has been observed in rodent and primate species [4–6].

Germ cell differentiation occurs in close interaction with testicular somatic cells, particularly Sertoli cells in the seminiferous epithelium of the testis. These SSCs reside in a particular microenvironment, the stem cell niche. In this niche, Sertoli cells, and presumably also cells from the interstitial compartment, modulate SSCs renewal and differentiation. The existence of a testicular stem cell niche is well documented but its cellular and molecular components are poorly understood and

I. Dobrinski, DVM, M.V.Sc., Ph.D. (✉)
Department of Comparative Biology and Experimental Medicine, University of Calgary,
3300 Hospital Drive, Calgary, AB T2N 4N1, Canada
e-mail: idobrins@ucalgary.ca

E. Seli and A. Agarwal (eds.), *Fertility Preservation in Males: Emerging Technologies and Clinical Applications*, DOI 10.1007/978-1-4614-5620-9_9,
© Springer Science+Business Media New York 2012

might vary between species [7–9]. However, transplantation of germ cells from a large variety of species into the mouse testis revealed that stem cell recognition and niche colonization are highly conserved among different species [10–13]. The plasticity of the niche to receive and host other germ line cells has been demonstrated; in addition to SSCs, primordial germ cells or teratocarcinoma cells have the potential to enter the niche and generate spermatogenesis [14, 15].

The SSC niche is established during the remodeling of the seminiferous epithelium that occurs during the postnatal phase of testis development. After an initial period of Sertoli cell proliferation, the seminiferous cords transform into tubules, and germ cells actively divide and differentiate to give rise to the first wave of spermatogenesis [16]. This first spermatogenic wave sets the framework for the future sperm production in adult life [2].

In primate adult testes, three types of spermatogonia are distinguished by morphological criteria: A_{dark}, A_{pale} and B spermatogonia [17–19]. The number of subsequent divisions of B-spermatogonia differ between primate species. For example, one division is described in men and four divisions (B1-B4) in macaques. Various models for spermatogonial kinetics have been described and are currently under debate signifying that the exact details of spermatogonial turnover in the primate testis are still largely unresolved [20, 21]. However, it is generally agreed that A_{dark} spermatogonia are mitotically quiescent and act as reserve stem cells since they become proliferatively active during pubertal expansion [22] and following depletion of spermatogonia due to irradiation or toxic exposure [23, 24]. On the other hand, A_{pale} spermatogonia proliferate regularly and are considered self-renewing progenitors [4, 5].

Rapidly dividing spermatogonia are highly sensitive to irradiation and toxins in adult and immature monkeys [24, 25]. Low doses of cytotoxic drugs or irradiation deplete the differentiating spermatogonia while less sensitive SSCs as well as spermatocytes and spermatids survive. Recovery of spermatogenesis occurs from the remaining SSCs and relies on the type, dose and fractionation of cytotoxic drugs and irradiation [23]. During recovery, testicular histology reveals an all-or-nothing pattern with areas of full spermatogenesis and areas with a Sertoli-cell-only pattern. This histological pattern during spermatogenic recovery indicates a critical role of SSCs for re-initiation of spermatogenesis and shows that with doses used in these primate studies the somatic environment is not heavily affected by chemotherapy or radiation exposure [26].

Cytotoxic drugs and irradiation are commonly used as therapeutic alternatives in oncological patients, and due to their toxicity for germ cells, development of alternatives that avoid testicular damage is highly desirable. Several strategies for protection of spermatogonial cells in these patients have been developed for application in a clinical setting (for review see Jahnukainen et al. [27]; Schlatt et al. [26]). While cryopreservation of sperm offers a standardized and routine option for fertility preservation in adults and pubertal patients, prepubertal children cannot donate spermatozoa for cryostorage. Moreover, cryopreserved sperm represent a finite source of gametes. Several alternatives have been discussed based on the high regenerative potential of the seminiferous epithelium that is supported by SSCs. Although the somatic environment can also be affected by irradiation and toxic exposures, this appears to be more resistant than germ cells [28]. Therefore, autologous transplantation of SSCs could provide an option for fertility preservation for oncological patients, regardless of their age [27].

Another alternative is xenografting of testicular tissue into immunodeficient mice. Prior to puberty, when the testis consists of seminiferous cords and the only germ cells present are spermatogonia, the developing testis appears to be more tolerant to transient ischemia making it possible to manipulate cells and tissue fragments while maintaining their full developmental potential. Since the first report in 2002, it has been demonstrated that testis tissue from a wide variety of larger species can initiate and complete development after transplantation into immunodeficient adult-castrated mice, and that the resulting sperm are capable of fertilizing and triggering normal embryo development [29–32]. This approach has been explored in primates, including humans, as a potential alternative for fertility preservation in prepubertal boys undergoing cytotoxic treatment during oncological processes [33–36]. It is the aim of this chapter to initially summarize the advances achieved in cryopreservation and ectopic xenografting of testis tissue as an alternative for fertility preservation, with emphasis in primates. Second, this chapter focuses on the methodology involved in both cryopreservation and xenotransplantation of testis tissue xenografting.

Testis Tissue Xenografting

Because of the architecture of its vascular and duct systems and the complexity of the seminiferous epithelium, the testis does not appear to be a suitable tissue for grafting. However, transplantation of testicular tissue has been performed for two centuries, and has provided important insights into testicular function. Autologous and homologous transplantation of testicular tissue has been reviewed by Gosden and Aubard [37, 38]. The generation of immunodeficient lines of mice allowed xenotransplantation of testicular and other tissues from large animals into rodent host [39]. Xenotransplantation of testicular tissue into immunodeficient mice (human fetal testis into the abdominal wall of adult nude mice) was first performed in 1974, resulting in initial development of the donor tissue [40]. Subsequently, Hochereau-de-Reviers and Perreau [41] transplanted ovine fetal testis into the scrotum of intact nude mice and reported differentiation of gonocytes into spermatogonia and primary spermatocytes. Complete cross species spermatogenesis was first reported in 2002 [29]. In that report, pieces of testis tissue from newborn pigs and goats were able to survive and displayed complete development with production of sperm.

These studies made clear that the immature testis was capable of initiating development if exposed to adult levels of gonadotropins, as lack of testis development during the prepubertal phase is largely due to low gonadotropin support. This immediate response of the testis tissue to initiate development results in an apparent shortening of the time required until first appearance of testicular sperm in immature testicular xenografts from species such as pigs and monkeys [29, 33]. In other species such as cats and cattle, this shortening is not evident mainly due to a developmental arrest of the tissue as grafts undergo meiosis [42–44].

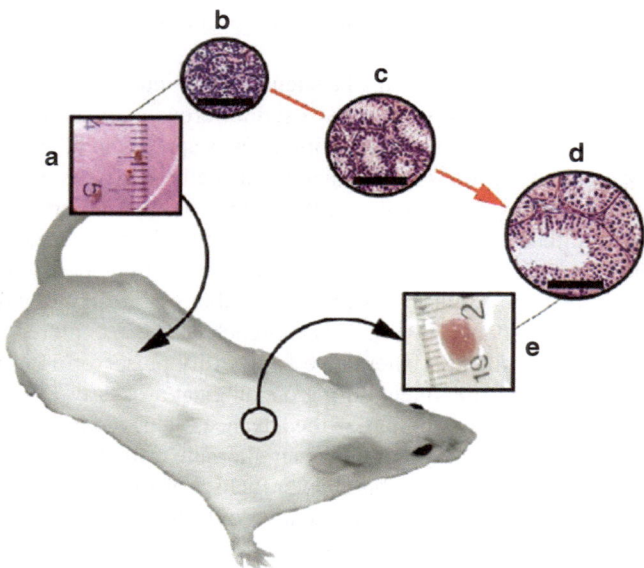

Fig. 9.1 Ectopic xenografting of immature testicular tissue from large animals into immunodeficient mice. Testis fragments (~1 mm³) from immature donors are transplanted under the back skin of immunodeficient mice (**a**), in this case a SCID mouse. Once transplanted, the testis fragments are able to survive and respond to mouse gonadotropins. The tissue develops from its immature stage containing typical seminiferous cords and gonocytes (**b**, **c**) to a mature-like stage, including the formation of fertilization competent sperm (**d**). Then, testis xenografts are collected for analysis or harvesting sperm (**e**). Bars = 50 μm

Once testis tissue has been transplanted in host mice, they rely on the re-establishment of the blood flow to survive and function [45]. The re-established blood supply provides murine LH and FSH to the xenografts. Grafts respond to this stimulation by recovering their endocrine function and establishing a feedback loop restoring the hypothalamic-pituitary-gonadal axis. Graft Leydig cells secrete testosterone in response to host LH, which along with recipient FSH induce the entire development of graft seminiferous cords similar to pubertal development (Fig. 9.1) [29, 31]. Graft testosterone supports spermatogenesis and maintains the integrity of the recipient reproductive tract; size and histological appearance of seminal vesicles, an androgen-dependent secondary sex gland in recipient mice, are indistinguishable from those of noncastrated mice [29, 31, 46].

Testis tissue xenografting has been shown to result in spermatogenesis in all domestic species evaluated, although at variable efficiency (reviewed by Rodriguez-Sosa and Dobrinski [47]). From the studies performed so far, one important factor affecting the survival and function of testis xenografts is the age or developmental stage of the donor. It is clear that use of tissue from prepubertal males leads to better survival and spermatogenesis outcome than that of pubertal or adult donors, and the capability of survival and development decreases markedly as the donor tissue has undergone meiosis [31, 44, 48–50].

Cryopreservation and Xenografting of Testicular Tissue for Fertility Preservation in Humans

An important finding that originated from the testis xenografting studies carried out in domestic species was the fertilizing capacity of the resulting sperm. This capacity was initially explored using intracytoplasmic sperm injection (ICSI). Injection of sperm from goat and pig xenografts into mouse oocytes resulted in initial stages of embryo development [29]. Similarly, when the procedure was performed with fresh or cryopreserved sperm from neonatal mouse testis xenografts, normal embryos were obtained, which generated normal offspring after being transferred into pseudo-pregnant females [32]. The potential to generate fertile sperm from prepubertal testes created interest to explore primate testis xenografting as a clinically relevant strategy for fertility preservation in boys. To explore this option, sperm obtained from testis xenografts of rhesus macaques were injected into rhesus oocytes, leading to the generation of morphologically normal embryos at the morulae and blastocyst stage [33]. Clinical use of cryopreservation and xenografting of testis tissue for fertility preservation in boys undergoing cancer treatment has been postulated. In this approach (Fig. 9.2), it is envisioned that prior to treatment, biopsies from reached maturity, has recovered from the original disease but has been rendered infertile by cytotoxic treatment, the tissue could be thawed and subsequently grafted into immunodeficient mice. Once the testis tissue has fully developed, grafts can be collected to harvest sperm to be used in ICSI to produce embryos for transfer. One advantage of this approach over autologous SSCs transplantation is the possibility to avoid the risk of reintroducing cancer cells into the patient testes that may be present in the donor cell suspension at the moment of collection [27, 51]. However, the possibility of host pathogens having an effect on the developing testis tissue and its resulting sperm must be considered in the case of testis tissue xenografting [27]. Therefore, use of this technique for fertility preservation in humans is still hypothetical and rigorous studies need to address ethical and safety issues before testis xenografting can be considered a feasible and safe option for fertility preservation in boys.

Recent Developments in Cryopreservation and Xenografting of Testicular Tissue in Primates

Ectopic Xenografting

One of the first reports of testis tissue xenografting from a large species explored the developmental potential of marmoset testis. Testis xenografts of immature marmosets showed a blockade in early germ cell differentiation [31]. Initially, it was not clear whether this was a species-specific event or an inherent problem with primate testis tissue. Subsequent studies showed that testis tissue from infant rhesus monkeys

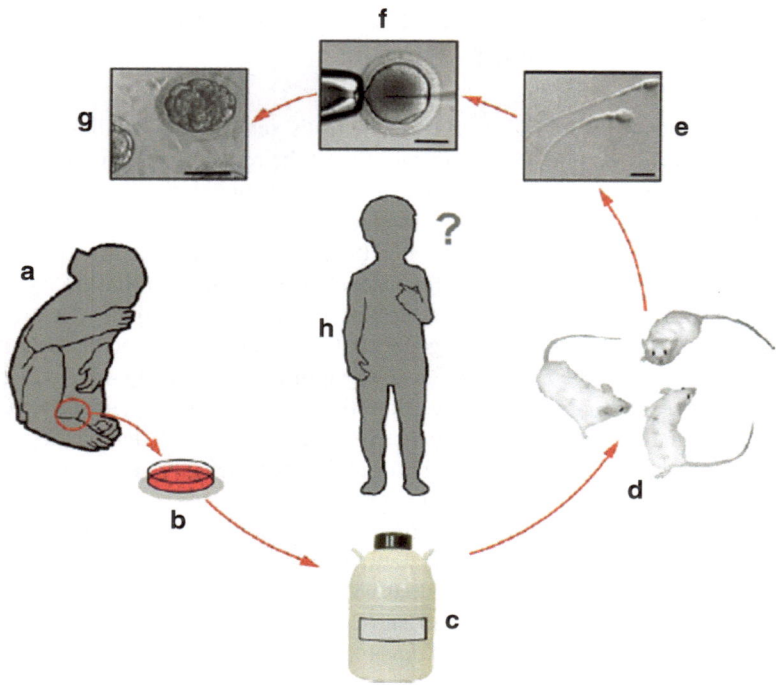

Fig. 9.2 Potential applications of testicular tissue xenografting for fertility preservation. Under this approach, testicular biopsies from immature males (**a**) are obtained and processed to produce fragments of suitable size for transplantation (**b**). The pieces are frozen and stored in Liquid nitrogen (**c**). The fragments are thawed at a later point and ectopically grafted into immunodeficient mice (**d**), where they are able to mature. Once tissue has matured, grafts are recovered for sperm harvesting (**e**), which is subsequently used for microinjection of oocytes (**f**) to generate embryos (**g**) ready to be transferred. In the case illustrated here, a rhesus monkey serves as a donor since it has been a suitable model to start establishing these techniques in humans (**h**). However, feasible and safe application of testis tissue xenografting for human fertility preservation has not yet been achieved. Bar = 10 μm in (**e**), and 50 μm in (**f, g**)

showed full developmental potential when transplanted to immunodeficient mice [33], and that gonadotropin supplementation promoted its maturation [52]. It is now known that the inability of marmoset testis tissue to develop in mice is due to an insensitivity to mouse LH caused by a deletion in exon 10 of the luteinizing hormone-receptor gene [53] and a hypersensitivity to the increased temperature at the subcutaneous site [54, 55]. Xenografting of human testis tissue has also been explored. The studies performed so far with human tissue show that the immature testis is more suitable for transplantation and differentiates once it is transplanted into host mice. However, so far complete spermatogenesis using immature human testis as xenografts has not been accomplished. Table 9.1 summarizes the strategies and outcome of studies which were performed using testicular xenografting from primate tissue donors.

Table 9.1 Outcome of testicular xenografting using primate testis tissue

Species	Donor age and developmental status	Cryopreservation	Recipient treatment	Collection time	Main outcome	References
Rhesus monkey	3 and 6 months. seminiferous cords	No	(A) PMSG, 10 IU twice a week (B) hCG, PMSG, 10 IU twice a week (C) None	7 months	Differentiation of cords into tubules relied on hormone treatment, and no difference was noticed between PMSG and hCG treatments. At collection, treated xenografts of 6-month-old donors showed mature sperm, whereas those of 3-month-old donors contained spermatocytes as the most advanced type of germ cells	[52]
Human	10–11 years, seminiferous cords/tubules containing spermatogonia	No	None	4 and 9 months	No difference was observed between xenografts of both collection points. No active spermatogenesis was obtained, and spermatogonia were the most advanced germ cell type	[62]
Human	7, 12 and 14 years, seminiferous cords/ tubules with spermatogonia (7, 12 and 14 years) and some tubules with focal spermatogenesis (14 years)[a]	Yes	None	6 months	Spermatocytes and spermatids were detected as the most advanced germ cells. No sperm with normal morphology were observed, but xenografts from the 14-year-old donor contained some sperm-like cells	[36]
Rhesus monkey	3, 6, 11 and 12 years, full spermatogenesis in 67–87% of tubules	No	None	<3, 3–6, and >6 months	Mostly atrophic tissue in xenografts from 11- and 12-year-old donors, very few tubules with spermatocytes in xenografts from the 6-year-old donor, and spermatocytes and spermatids in grafts of the 3-year-old donor at <3, 3–6 months	[50]

(continued)

Table 9.1 (continued)

Species	Donor age and developmental status	Cryopreservation	Recipient treatment	Collection time	Main outcome	References
Rhesus monkey	16 and 19 months, seminiferous cords with spermatogonia	No	None, but testicular fragments were exposed to irradiation (0, 0.5, 1, and 4 Gy)	4 months	Type B spermatogonia were the most advanced germ cells present in all xenografts. However, spermatogonial numbers were reduced after exposure of donor tissue fragments to 0.5 Gy of irradiation. Spermatogonia were almost depleted in xenografts exposed to 1 and 4 Gy	[56]
Human	2–12 years, cryptorchid, seminiferous cords with spermatogonia[a]	Yes	None	21 days	Spermatogonia were the most advanced type of germ cells. In comparison to donor tissue the total number of spermatogonia was decreased in xenografts. However, number of proliferating spermatogonia was higher in xenografts than in donor tissue	[35]
Rhesus monkey	18 and 21 months, seminiferous cords with spermatogonia	Yes	None	3 and 5 months	Type B spermatogonia and spermatocytes were the most advanced type of germ cells at 3 and 5 months, respectively	[34]
Human	Fetal, 20 and 26 weeks of gestation, seminiferous cords	No	None	116 and 135 days	Differentiation of cords into tubules was evident; increase in tubule diameter, initial stages lumen formation, and spermatogonia located on the basal lamina were observed	[63]

Rhesus monkey	16 and 19 months, seminiferous cords with spermatogonia	No	(A) None (B) Busulfan (38 mg/kg) in saline and DMSO (1:1) at 7 months (C) Saline and DMSO (1:1) at 7 months	7 months (A), and 8 months (B and C)	At 7 months (Group A), type B spermatogonia and spermatocytes were the most advanced germ cells. At 8 months, in vehicle-treated xenografts (Group B) round spermatids were detected, whereas in Busulfan-treated grafts no spermatogenesis was observed, and a decrease in the number of tubules with germ cells was evident	[25]
Human	Adult, complete spermatogenesis	No	None	30–195 days	Most xenografts (~60%) showed sclerosis. However, in ~22% of them spermatogonia were still observed, with increasing sclerosis in time	[48]
Human	Adult, intact and disturbed spermatogenesis	No	None	2–19 weeks	Xenografts from donors with obstructive azoospermia, hypospermatogenesis, and Sertoli cells only (SCO) showed complete atrophy and hyalinization, with no germ cells. Those from donors with testicular cancer showed SCO tubules and some with type A spermatogonia. Xenografts from transsexual donors (post estradiol treatment) contained some A- and few B-spermatogonia	[64]

(continued)

Table 9.1 (continued)

Species	Donor age and developmental status	Cryopreservation	Recipient treatment	Collection time	Main outcome	References
Rhesus monkey	13 months, seminiferous cords	No	None	2–7 months	Increase in cord diameter and spermatogonia located on the basal lamina were evident at 2 months. Transformation of cords into tubules, with presence of spermatocytes and round spermatids were observed at 4 months. Complete spermatogenesis was observed at 7 months	[33]
Marmoset	Newborn, and 1,3, and 7 months, seminiferous cords	No	None, but tissue was co-engrafted with hamster tissue	3 months	Spermatogonia were the most advanced type of germ cells in marmoset xenografts, and co-engraftment with hamster tissue did not induce any further differentiation of marmoset germ cells	[53]
Marmoset	Newborn, seminiferous cords	No	Gonadotropins (Pergonal, three injections of 1.5 IU/week) from day 100 to 135	100 and 135 days	Spermatogonia and spermatocytes were the most advanced types of germ cells at 100 and 13 days, respectively	[31]

[a]Xenografting was performed into the scrotum of castrated host mice

Besides providing a potential alternative for fertility preservation, xenografting of testis tissue is an elegant experimental strategy to evaluate the effects of hormonal manipulation or exposure to radiation or gonadotoxins on testicular development. The gonadotoxic effects of busulfan were reproduced in monkey testis xenografts [25]. Similarly, the gonadotoxic effect of irradiation was shown when testis tissue from rhesus monkeys was exposed to irradiation doses and subsequently transplanted in mice [56]. Stimulation of grafted infant primate testis tissue with exogenous gonadotropins supported its maturation, and this was due to a direct effect on Sertoli cells [52]. This illustrated the potential use of exogenous hormonal supplement for manipulation of development of testis xenografts. As experimental strategy, an advantage of xenografting is the small number of donors needed to perform valid comparative studies. Instead of exposing groups of monkeys to different treatments, groups of mice carrying xenografts from few juvenile monkeys can be exposed to different regimens.

Although in the current chapter we have mainly focused on xenotransplantation of testis tissue, it is worth mentioning that autologous transplantation of primate testis tissue has also been performed. Autologous transplantation of testis tissue was performed in newborn marmosets to explore whether the developmental blockade in the mouse host was only due to the hormonal milieu provided by the rodent host.

Partial differentiation of the transplanted tissue was obtained, with spermatocytes as the most advanced type of germ cell. This was attributed to the hyperthermia of the ectopic site [54]. Recently, testis tissue from immature and adult marmosets was transplanted in autologous recipients. Unlike adult tissue, immature tissue showed good survival and recovery, and similar to the previous study, ectopic placement of grafts resulted in partial differentiation, with spermatocytes as the most advanced germ cells. However, when the tissue was transplanted into the scrotum, mature spermatids were obtained [55]. Finally, partial differentiation of immature testis tissue after autologous transplantation and gonadotropin supplementation has also been reported in rhesus monkeys [57].

Cryopreservation

Long-term preservation of testicular tissue by cryostorage is an essential prerequisite for testis tissue xenografting to be widely applicable to clinical settings. The original studies performed in pigs and goats showed that freezing and thawing of testis tissue prior to transplantation did not affect its capability to undergo full development in the host mice [29]. Moreover, in vitro microinsemination using sperm harvested from frozen and thawed rabbit testis tissue orthotopically transplanted in noncastrated mice resulted in embryos and apparently normal offspring [31]. Since then, cryopreservation of testicular tissue has been evaluated in several species with varying degrees of germ cell differentiation after thawing and grafting (Table 9.2). In humans and other primates, the most promising results have been obtained with DMSO as the cryoprotectant component of the freezing medium. The structural attributes of human testicular tissue were maintained by freezing with a programmable

Table 9.2 Testicular cryopreservation in non-rodent species

Species	Freezing method	Xenografting	Outcome	References
Human	0.7 M DMSO HBSS, supplemented with sucrose and human serum albumin, automated freezing	Yes[a]	Testis tissue showed minor degenerative signs. Incomplete spermatogenesis, with round spermatids as most advanced type of germ cells, with presence of few abnormal sperm-like cells in the xenografts of one donor	[36]
Human	(A) HBSS plus 5% DMSO and 5% human serum albumin. Automated freezing at cooling rate 1°C/min to 0°C, 0.5°C/min to −40°C, 7°C to −70°C with final plunging into liquid nitrogen (B) Same medium as in (A), Automated freezing at cooling rate 1°C/min to −8°C, 10°C/min to −80°C with final plunging into liquid nitrogen	No	After thawing, no structural differences were noticed between frozen and fresh tissue. Best method: B	[59]
Human	(A) Leibovitz L-15 plus 1.5 M ethylene glycol, 0.1 M sucrose and 0.1% human serum albumin, automated freezing	No	After thawing, no structural and functional (in vitro culture) differences were noticed between frozen and fresh tissue.	[65]
	(B) Same as in A, except that Leibovitz medium was replaced by PBS		No difference was observed between the freezing methods used	
Human	(A) Egg yolk-based medium containing 12% glycerol, supplemented with sucrose and patient serum, automated freezing	No	After thawing, no structural differences were noticed between frozen and fresh tissue. Best method: C	[58]
	(B) 1.5 M 1,2-propanediol HBSS supplemented with sucrose and patient serum, automated freezing			
	(C) 0.7 M DMSO HBSS, supplemented with sucrose and patient serum, automated freezing			
Rhesus monkey	(A) 1.4 M ethylene glycol Leibovitz L-15 plus 10% FCS, automated freezing	Yes	Initiation of spermatogenesis, with spermatocytes as the most advanced type of germ cells. Best method: B	[34]
	(B) 1.4 M DMSO Leibovitz L-15 plus 10% FCS, automated freezing			
	(C) Leibovitz L-15 plus 10% FCS, automated freezing			

Pig	(A) DMSO, DMEM, and FCS at 1:3:1 ratio, conventional slow-freezing in alcohol bath (B) Leibovitz L-15 plus 2% FCS and 10% DMSO, automated freezing (C) Vitrification with ethylene glycol containing 0.9% NaCl and 0.5 M raffinose	Yes	At collection spermatogenesis was delayed in comparison to fresh tissue xenografts, with spermatocytes as the most advanced type of germ cells. Best method: A	[60]
Pig	DMSO, DMEM, and FCS at 1:3:1 ratio, and conventional slow-freezing in alcohol bath	Yes	Complete spermatogenesis	[29]
Goat	DMSO, DMEM, and FCS at 1:3:1 ratio, and conventional slow-freezing	Yes	Complete spermatogenesis	[29]
Rabbit	DMSO, DMEM, and FCS, automated freezing	Yes[b]	Complete spermatogenesis. Normal progeny was obtained by ICSI with resulting sperm	[66]

[a]Xenografting was performed into the scrotum of castrated host mice
[b]Recipient mice were not castrated and donor testis tissue was transplanted into the testis

freezer using 0.7 M DMSO. However, xenografting of the frozen tissue was not attempted [58, 59]. In the monkey study, 1.4 M DMSO resulted in better spermatogenesis outcome in comparison to 0.7 M DMSO [34]. More recently, immature testicular tissue frozen with 0.7 M DMSO and subsequently thawed and transplanted into the scrotum of castrated nude mice was able to differentiate. However, most of the xenografts arrested at the meiotic stage and except for few morphologically abnormal sperm in the samples of one donor, no normal sperm was obtained [36]. Importantly, studies performed in rhesus monkeys and pigs demonstrated that similarity of frozen-thawed and fresh tissue does not necessarily translate to equivalent development after grafting into mouse hosts [34, 60].

A complementary alternative to freezing testicular tissue is the possibility of freezing sperm that are harvested from testis xenografts [61]. Snap-frozen sperm harvested from mouse allografts led to normal embryos and progeny when microinjected into mouse oocytes [32]. Similarly, microinjection with snap-frozen sperm from pig testis xenografts resulted in generation of morphologically normal embryos, albeit of a lower efficiency in comparison to testicular, epididymal, or ejaculated sperm [30]. Freezing sperm recovered from pig testis xenografts resulted in viable sperm after thawing, but the number of viable sperm was reduced in comparison to frozen-thawed testicular sperm [60]. Differences in the characteristics or fertilizing ability of the sperm obtained from frozenthawed testis xenografts has been attributed, at least in part, to the senescense of the resulting sperm. Since testis xenografts lack the excurrent ducts of in situ testes, resulting sperm accumulate in the lumen of the seminiferous tubules. Since so far there is no way to distinguish recently-formed sperm from those that have been accumulated, old sperm could be employed in ICSI leading to decreased fertilization outcome [30, 60]. This emphasizes the need for developing efficient methods not only for testis tissue and sperm cryopreservation, but also to determine optimal times for xenograft recovery and sperm harvesting. See Chap. 14 and [61] for a detailed protocol for xenografting.

Conclusions

Based on results obtained from studies in domestic species and primates, testicular tissue xenografting represents a promising alternative for fertility preservation in prepubertal boys. Cryostorage of the patient testis tissue is essential for testis tissue xenografting to become a practical alternative under clinical settings. Important advances have been made towards that goal. However, additional studies need to be performed to address ethical and safety issues before testis tissue xenografting can be used in humans. So far, complete spermatogenesis has not been obtained from cryopreserved human testis tissue transplanted into immunodeficient mice. A key area to address in the future is the development of freezing methods that preserve not only the structure and developmental potential of prepubertal human tissue, but also the fertilization capability of the resulting sperm.

See Chap. 14

References

1. Russell LD, Ettlin RA, SinhaHikim AP, et al. Mammalian spermatogenesis. In: Russell LD, Ettlin RA, SinhaHikim AP, Clegg ED, editors. Histological and histopathological evaluation of the testis. St. Louis: Cache River Press; 1990. p. 1–38.
2. de Rooij DG, Russell LD. All you wanted to know about spermatogonia but were afraid to ask. J Androl. 2000;21:776–98.
3. Luo J, Megee S, Dobrinski I. Asymmetric distribution of UCH-L1 in spermatogonia is associated with maintenance and differentiation of spermatogonial stem cells. J Cell Physiol. 2009;220:460–8.
4. Ehmcke J, Luetjens CM, Schlatt S. Clonal organization of proliferating spermatogonial stem cells in adult males of two species of non-human primates, Macaca mulatta and Callithrix jacchus. Biol Reprod. 2005;72:293–300.
5. Ehmcke J, Simorangkir DR, Schlatt S. Identification of the starting point for spermatogenesis and characterization of the testicular stem cell in adult male rhesus monkeys. Hum Reprod. 2005;20:1185–93.
6. Nakagawa T, Sharma M, Nabeshima Y, et al. Functional hierarchy and reversibility within the murine spermatogenic stem cell compartment. Science. 2010;328:62–7.
7. Orwig KE, Ryu BY, Master SR, et al. Genes involved in post-transcriptional regulation are overrepresented in stem/progenitor spermatogonia of cryptorchid mouse testes. Stem Cells. 2008;26:927–38.
8. Kostereva N, Hofmann MC. Regulation of the spermatogonial stem cell niche. Reprod Domest Anim. 2009;43(Suppl 2):386–92.
9. Oatley JM, Oatley MJ, Avarbock MR, et al. Colony stimulating factor 1 is an extrinsic stimulator of mouse spermatogonial stem cell self-renewal. Development. 2009;136:1191–9.
10. Dobrinski I, Avarbock MR, Brinster RL. Transplantation of germ cells from rabbit and dogs into mouse testes. Biol Reprod. 1999;61:1331–9.
11. Dobrinski I, Avarbock MR, Brinster RL. Germ cell transplantation from large domestic animals into mouse testes. Mol Reprod Dev. 2000;57:270–9.
12. Nagano N, McCarrey JR, Brinster RL. Primate spermatogonial stem cells colonize mouse testis. Biol Reprod. 2001;64:1409–16.
13. Nagano N, Patrizio P, Brinster RL. Long-term survival of human spermatogonial stem cells in mouse testes. Fertil Steril. 2002;78:1225–33.
14. Nayernia K, Li M, Jaroszynski L, Khusainov R, et al. Stem cell based therapeutical approach of male infertility by teratocarcinoma derived germ cells. Hum Mol Genet. 2004;13:1451–60.
15. Chuma S, Kanatsu-Shinohara M, Inoue K, et al. Spermatogenesis from epiblast and primordial germ cells following transplantation into postnatal mouse testis. Development. 2005;132:117–22.
16. Orth JM. Cell biology of testicular development in the fetus and neonate. In: Desjardins C, Ewing LL, editors. Cell and molecular biology of the testis. New York: Oxford University Press; 1993. p. 3–42.
17. Clermont Y, Leblond CP. Differentiation and renewal of spermatogonia in the monkey, *Macaca rhesus*. Am J Anat. 1959;104:237–73.
18. Clermont Y. Spermatogenesis in man. A study of the spermatogonial population. Fertil Steril. 1966;17:705–21.
19. Clermont Y. Two classes of spermatogonial stem cells in the monkey *(Cercopithecus aethiops)*. Am J Anat. 1969;126:57–71.
20. Ehmcke J, Schlatt S. A revised model for spermatogonial expansion in man: lessons from non-human primates. Reproduction. 2006;132:673–80.
21. Amann RP. The cycle of the seminiferous epithelium in humans: a need to revisit? J Androl. 2008;29:469–87.
22. Simorangkir DR, Marshall GR, Ehmcke J, et al. Prepubertal expansion of dark and pale type A spermatogonia in the rhesus monkey *(Macaca mulatta)* results from proliferation during

infantile and juvenile development in a relatively gonadotropin independent manner. Biol Reprod. 2005;73:1109–15.

23. van Alphen MMA, van den Kant HJG, de Rooij DG. Repopulation of the seminiferous epithelium of the rhesus monkey after X-irradiation. Radiat Res. 1988;113:487–500.

24. van Alphen MMA, van den Kant HJG, Davids JAG, et al. Dose–response studies on the spermatogonial stem cells of the Rhesus monkey (*Macaca mulatta*) after X-irradiation. Radiat Res. 1989;199:443–51.

25. Jahnukainen K, Ehmcke J, Schlatt S. Testicular xenografts: a novel approach to study cytotoxic damage in juvenile primate testis. Cancer Res. 2006;66:3813–8.

26. Schlatt S, Ehmcke J, Jahnukainen K. Testicular stem cells for fertility preservations: preclinical studies on male germ cell transplantation and testicular grafting. Pediatr Blood Cancer. 2009;53:274–80.

27. Jahnukainen K, Ehmcke J, Soder O, et al. Clinical potential and putative risks of fertility preservation in children utilizing gonadal tissue or germ line stem cells. Pediatr Res. 2006;59:40R–7.

28. Zhang Z, Shao S, Meistrich ML. Irradiated mouse testes efficiently support spermatogenesis derived from donor germ cells of mice and rats. J Androl. 2006;27:365–75.

29. Honaramooz A, Snedaker A, Bioani M, et al. Sperm from neonatal testes grafted in mice. Nature. 2002;418:778–81.

30. Honaramooz A, Cui X, Kim N, et al. Porcine embryos produced after intracytoplasmic sperm injection using xenogeneic pig sperm from neonatal testis tissue grafted in mice. Reprod Fertil Dev. 2008;20:802–7.

31. Schlatt S, Kim SS, Gosden R. Spermatogenesis and steroidogenesis in mouse, hamster and monkey testicular tissue after cryopreservation and heterotopic grafting to castrated host. Reproduction. 2002;124:339–46.

32. Schlatt S, Honaramooz A, Bioani M, et al. Progeny from sperm obtained after ectopic grafting of neonatal mouse testes. Biol Reprod. 2003;68:2331–5.

33. Honaramooz A, Li M, Penedo CT, et al. Accelerated maturation of primate testis by xenografting into mice. Biol Reprod. 2004;70:1500–3.

34. Jahnukainen K, Ehmcke J, Hergenrother SD, et al. Effect of cold storage and cryopreservation of immature non-human primate testicular tissue on spermatogonial stem cell potential in xenografts. Hum Reprod. 2007;22:1060–7.

35. Wyns C, Curaba M, Martinez-Madrid B, et al. Spermatogonial survival after cryopreservation and short-term orthotopic immature human cryptorchid testicular tissue grafting to immunodeficient mice. Hum Reprod. 2007;22:1603–11.

36. Wyns C, Van Langendonckt A, Wese FX, et al. Longterm spermatogonial survival in cryopreserved and xenografted immature human testicular tissue. Hum Reprod. 2008;23:2402–14.

37. Gosden RG, Aubard Y. Why transplant gonadal tissue? In: Gosden RG, Aubard Y, editors. Transplantation of ovarian and testicular tissues. Austin: Landes/Chapman & Hall; 1996. p. 1–15.

38. Gosden RG, Aubard Y. Transplantation of testicular tissue. In: Gosden RG, Aubard Y, editors. Transplantation of ovarian and testicular tissues. Austin: Landes/Chapman & Hall; 1996. p. 89–97.

39. Paris MCJ, Snow M, Cox S, et al. Xenotransplantation: a tool for reproductive biology and animal conservation? Theriogenology. 2004;61:277–91.

40. Skakkebaek NE, Jensen G, Povlsen CO, et al. Heterotransplantation of human foetal testicular and ovarian tissue to the mouse mutant nude. Acta Obstet Gynecol Scand. 1974;53:73–5.

41. Hochereau-de-Reviers MT, Perreau C. Induced differentiation of ovine foetal gonocytes after grafting in the scrotum of nude mice. Reprod Nutr Dev. 1997;37:469–76.

42. Rathi R, Honaramooz A, Zeng W, et al. Germ cell fate and seminiferous tubule development in bovine testis xenografts. Reproduction. 2005;130:923–9.

43. Oatley JM, Reeves JJ, McLean DJ. Establishment of spermatogenesis in neonatal bovine testicular tissue following ectopic xenografting varies with donor age. Biol Reprod. 2005;72:358–64.

44. Kim Y, Selvaraj V, Pukazhenthi B, et al. Effect of donor age on success of spermatogenesis in feline testis xenografts. Reprod Fertil Dev. 2007;19:869–76.
45. Schlatt S, Westernstroer B, Gassei K, et al. Donor-host involvement in immature rat testis xenografting into nude mouse hosts. Biol Reprod. 2010;82:888–95.
46. Rodriguez-Sosa JR, Foster RA, Hahnel A. Development of strips of ovine testes after xenografting under the skin of mice and co-transplantation of exogenous spermatogonia with grafts. Reproduction. 2010;139:227–35.
47. Rodriguez-Sosa JR, Dobrinski I. Recent developments in testis tissue xenografting. Reproduction. 2009;138:187–94.
48. Geens M, De Block G, Goossens E, et al. Spermatogonial survival after grafting human testicular tissue to immunodeficient mice. Hum Reprod. 2006;21:390–6.
49. Rathi R, Honaramooz A, Zeng W, et al. Germ cell development in equine testis tissue xenografted into mice. Reproduction. 2006;131:1091–8.
50. Arregui L, Rathi R, Zeng W, et al. Xenografting of adult mammalian testis tissue. Anim Reprod Sci. 2008;106:65–76.
51. Aslam I, Fishel S, Moore H, et al. Fertility preservation of boys undergoing anti-cancer therapy: a review of the existing situation and prospects for the future. Hum Reprod. 2000;15:2154–9.
52. Rathi R, Zeng W, Megee S, et al. Maturation of testicular tissue from infant monkeys after xenografting into mice. Endocrinology. 2008;149:5288–96.
53. Wistuba J, Mundry M, Luetjens CM, et al. Cografting of hamster *(Phodopus sungorus)* and marmoset *(Callithrix jacchus)* testicular tissues into nude mice does not overcome blockade of early spermatogenic differentiation in primate grafts. Biol Reprod. 2004;7:2087–91.
54. Wistuba J, Luetjens CM, Wesselmann R. Meiosis in autologous ectopic transplants of immature testicular tissue grafted to *Callithrix jacchus*. Biol Reprod. 2006;74:706–13.
55. Luetjens CM, Stukenborg JB, Nieschlag E, et al. Complete spermatogenesis in orthotopic but not in ectopic transplants of autologously grafted marmoset testicular tissue. Endocrinology. 2008;149:1736–47.
56. Jahnukainen K, Ehmcke J, Nurmio M. Irradiation causes acute and long-term spermatogonial depletion in cultured and xenotransplanted testicular tissue from juvenile nonhuman primates. Endocrinology. 2007;148:5541–8.
57. Orwig KE, Schlatt S. Cryopreservation and transplantation of spermatogonia and testis tissue for preservation of male fertility. J Natl Cancer Inst Monogr. 2005;34:56.
58. Keros V, Rosenlund B, Hultenby K, et al. Optimizing cryopreservation of human testicular tissue: comparison of protocols with glycerol, propanediol and dimethylsulphoxide as cryoprotectants. Hum Reprod. 2005;20:1676–87.
59. Keros V, Hultenby K, Borgstrom B, et al. Methods of cryopreservation of testicular tissue with viable spermatogonia in pre-pubertal boys undergoing gonadotoxic cancer treatment. Hum Reprod. 2007;22:1384–895.
60. Zeng W, Snedaker AK, Megee S, et al. Preservation and transplantation of porcine testis tissue. Reprod Fertil Dev. 2009;21:489–97.
61. Rathi R, Dobrinski I. Ectopic grafting of mammalian testis tissue into mouse hosts. In: Hou SX, Singh SR, editors. Methods in molecular biology. Germ line stem cells. Totowa: Humana Press; 2008. p. 139–48.
62. Goossens E, Geens M, De Block G, et al. Spermatogonial survival in long-term human prepubertal xenografts. Fertil Steril. 2008;90:2019–22.
63. Yu J, Cai ZM, Wan HJ, et al. Development of neonatal mouse and fetal human testicular tissue as ectopic grafts in immunodeficient mice. Asian J Androl. 2006;8:393–403.
64. Schlatt S, Honaramooz A, Ehmcke J, et al. Limited survival of adult human testicular tissue as ectopic xenograft. Hum Reprod. 2006;21:384–9.
65. Kvist K, Thorup J, Byskov AG, et al. Cryopreservation of intact testicular tissue from boys with cryptorchidism. Hum Reprod. 2006;21:484–91.
66. Shinohara T, Inoue K, Ogonuki N, et al. Birth of offspring following transplantation of cryopreserved immature testicular pieces and in-vitro microinsemination. Hum Reprod. 2002;17:3039–45.

Chapter 10
Stem Cells and Fertility Preservation in Males

Marcia Riboldi, Ana Isabel Marqués Marí, and Carlos Simón

Currently, around 70 million couples suffer because of infertility worldwide [1, 2]. Around 15% of couples are not able to conceive within 1 year and treatments for infertility are necessary because of different causes [3], where the male factor is responsible for 50% of infertile couples. In general, infertility disorders affect a large percentage of men, roughly 30–40% [4].

Approximately 30–40% of men are survivors of malignant diseases such as cancer, and the consequences of this are subfertility or infertility as secondary effects. One in 650 children develops cancer by the age of 15, and 50–60% will be cured. Today, 1 in 1,000 young adults is a childhood cancer survivor [5]. Testicular germ cell cancer (TGCC) and lymphoma are the most frequent cancer types in young patients. In TGCC, for example, the cure rate is around 95% [6]. However, infertility is a huge burden for these cancer survivors.

The leading cause of male infertility is azoospermia which can be attributed to a variable number of causes. Male genital tract obstructions might result from infections, previous inguinal, and scrotal surgery (vasectomy), secondary genital duct obstruction due to inflammation, failed vasectomy reversal, congenital bilateral absence of the *vas deferens* [7], among others. The prevalence of azoospermia is approximately 1% among all men [8], and ranges between 10% and 15% among infertile men [9]. Whereas spermatozoa are abundant in the epididymis in male patients suffering tract obstructions, allowing a high rate of retrieval success, only a few *foci* with spermatogenesis are found in testes in secretory azoospermia [10–13]. However, the worst prognosis for infertility is for those in which no germ cells (GCs) are present [14, 15].

Stem cells (SCs) are a specific population of undifferentiated cells that are capable of selfrenewal and differentiation into all the cell types of the organism. These cells were isolated for the first time in 1998 [16]. Thanks to their unique characteristics, they have opened a new horizon and have raised the possibility of their future use in regenerative

C. Simón, M.D., Ph.D. (✉)
Stem Cell Bank, Prince Felipe Research Centre, Autopista del Saler 16,
Camino de las Moreras, Valencia 46012, Spain

E. Seli and A. Agarwal (eds.), *Fertility Preservation in Males: Emerging Technologies
and Clinical Applications*, DOI 10.1007/978-1-4614-5620-9_10,
© Springer Science+Business Media New York 2012

medicine. Embryonic stem cells (ESC) are obtained from preimplantation stages of the embryo (blastocyst, morula, blastomere) [16–18] and have pluripotent characteristics since they are capable of differentiating into all cell types, including GCs. Adult SCs have been described in a wide range of tissues, including brain [19], blood [20, 21], bone marrow [22], fat [23] or skin [24, 25], among others, and also in testes [26]. These cells have pluripotent potential and can differentiate into various types of cells depending on their origin [27].

Male GCs develop into spermatogonial stem cells (SSCs) which are fundamental for the spermatogenesis process and have the ability to selfrenew and to generate differentiated GCs [28, 29].

Male Germline

GCs are responsible for the transmission of individual genetic information to the next generation, and it is through this process that they assure the continuity of species [30]. GCs are derived from a population of primordial germ cells (PGC) [31, 32] that arise from the proximal epiblast [31] to then migrate into primitive gonads and proliferate by increasing in number (Fig. 10.1).

In men, once PGCs arrive at the genital ridge, they are enclosed by somatic SCs and become gonocytes [33]. Cells proliferate for a few days and then arrest in the G0/G1 phase during fetal development. After birth, PGCs migrate to the basement membrane of the seminiferous tubules and become SSCs. SSCs divide asymmetrically to give rise to one stem cell, and to one spermatogonia that initiates its differentiation into spermatozoa [34, 35] (Fig. 10.2).

Maintenance of the self-renewal and differentiation of the germ SCs in the testis depends on the niche that is located at the basal membrane of the seminiferous tubules [36], where only 1% of cells are SSCs [37]. The SSC niche is maintained due to the many factors produced and secreted by SCs. Of these, one of the most important is the glial cell line-derived neurotrophic factor (GDNF) as it is essential for the self-renewal and survival of SSCs [38, 39].

The presence in testes of a stem cell population (SSCs) responsible for continuous sperm production was demonstrated in 1994 by Brinster and Zimmermann in mice [30]. In 2008, Conrad and collaborators reported the isolation of SSCs from adult testicular tissue after generating human adult germline stem cells (haGSCs). These

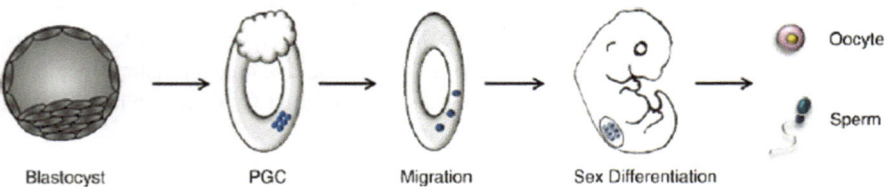

Fig. 10.1 PGC population and sex differentiation

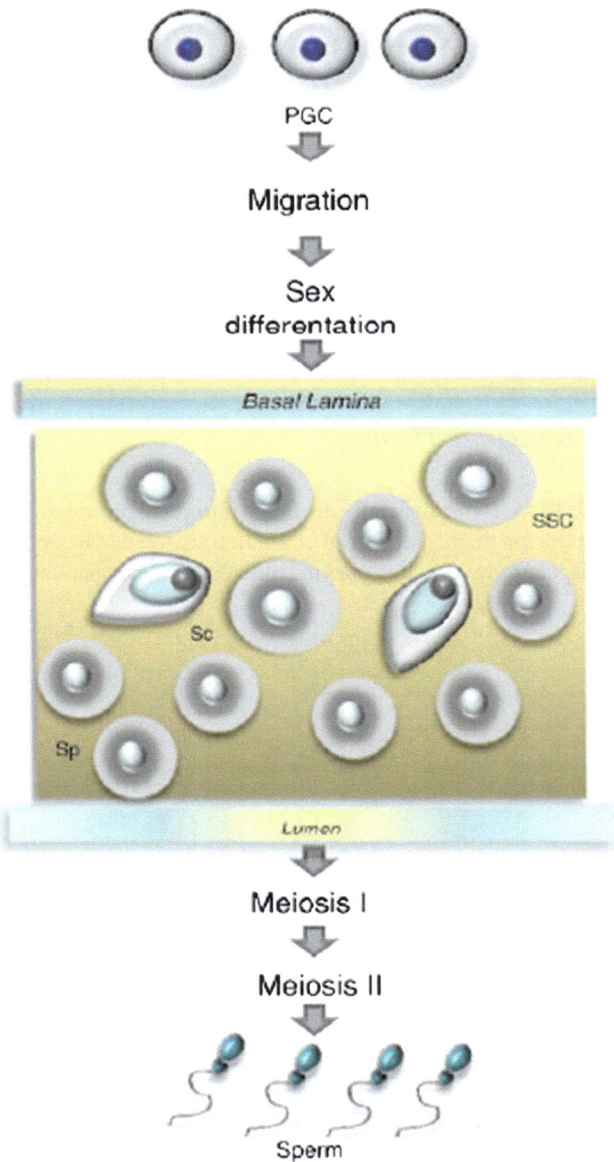

Fig. 10.2 Spermatogenesis. *SSC* spermatogonial stem cell, *Sc* Sertoli cell, *Sp* spermatogonial cell

cell lines showed similar cellular and molecular characteristics to ESC and GCs. These cells are pluripotent and capable of giving rise to many different cell types in vitro and of forming teratomas when injected into immunosuppressed mice [40]. Very recently, however, these findings have been questioned by Scholer's group which demonstrated that haGSCs and fibroblasts have a similar gene-expression profile. Therefore, the pluripotency of the human SSCs population is currently under debate [41].

Methods of Isolation and Culture of Germ Cells

The first requisite to establish the culture of SSCs is to start with pure cell populations, while the second is to procure optimal conditions to support the survival and expansion of cells. Various methods and protocols have been described for the isolation, culture, and maintenance of the SSCs in vitro (Fig. 10.3).

Digestion of Testicular Tissue

Initially, testicular tissue is mechanically disrupted into small pieces, and then enzymatic techniques are used for the dispersion of seminiferous tubules to obtain a single cell suspension. Collagenase was the first enzyme utilized for the dispersion of seminiferous

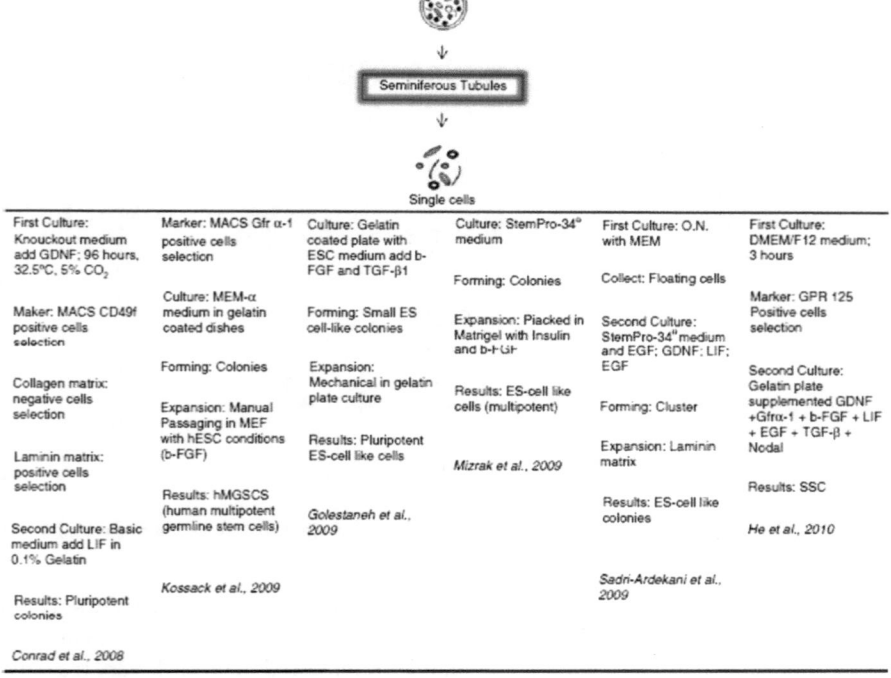

Fig. 10.3 Protocols for the isolation, culture and maintenance of the SSCs in vitro

tubules, together with DNaseI, hyaluronidase, and trypsin. Various combinations of enzymes and concentrations have been described for different groups.

Conrad et al. used 0.5 g/mL collagenase type VI and 0.25 g/mL dispase II for 30 min at 37°C [40]. Dym's group used 1 mg/mL collagenase IV and 2 μg/μL DNase I [42]. Reijo-Pera reported a first incubation of 30 min at 37°C in 10 mg/mL collagenase [43].

Other types of enzymes that used to obtain individual cells were 4 mg/mL collagenase IV, 2.5 mg/mL hyaluronidase, 2 mg/mL trypsin, and 1 μg/μL DNase I for Dym's group, and 1 mg/mL collagenase IV, 1.5 mg/mL hyaluronidase, and 1 mg/mL trypsin for Reijo-Pera [40, 42, 44]. After digestion, the supernatant is filtered through 40, 50, and 70 μm filters [40, 45], and the cells obtained are cultured.

Isolation of SSCs

Fluorescence-activated cell sorting (FACS) or magnetic-activated cell sorting (MACS) methods are coupled with a diverse number of markers to identify SSCs (Table 10.1). Extracellular matrices have been used for the cell purification process. Candidate cells are incubated for 4 h at 32.5°C onto a collagen matrix (5 mg cm^2). Then the supernatant is aspirated and incubated onto a laminin matrix (4.4 mg cm^2) for 45 min at 32.5°C [40]. This methodology was developed to specifically select

Table 10.1 Markers used to identify SSCs

Surface markers	Species	References
α6-integrin (CD49f)	Mouse	[55]
	Human	[40]
	Primate	[71]
β1-integrin	Mouse	[55]
Thy-1 (CD90)	Mouse	[47]
	Rat	[57]
	Primate	[56]
	Human	[40]
CD9	Mouse	[56]
	Rat	
Ep-CAM	Mouse	[57]
	Rat	
E-cadherin	Mouse	[58]
Gfrα1	Mouse	[59]
	Human	[60]
		[43]
CD24	Mouse	[72]
GPR125	Mouse	[61, 62]
	Human	[44]

Table 10.2 Different exogenous agents added to the culture medium to induce cellular processes

Agent	In vivo	In vitro	References
GDNF	Proliferation	Survival	[48, 57, 63–67]
	Undifferentiated	Proliferation	
	Maintenance	Self-renewal	
	Regulation	Expansion	
	Self-renewal	Proliferation	
Bfgf		Survival	[47]
		Proliferation	
		Expansion	
EGF	Proliferation	Survival	[48, 67]
		Proliferation	
		Still unclear	
IGF-I		Increase in number	[48]
LIF	Maturation	Survive	[67–69]
	Self-renewal	Proliferation	
		Still unclear	
GFRα-1		Expansion	[47]
		Proliferation	
SCF	Proliferation		[67, 70]
	Transition types spermatogonia		
FGF2	Regulation	Self-renewal	[67]
	Behavior	Differentiation	

the cell population attached to the laminin matrix [40, 46]. These cells showed the typical morphology of SSC, and loosely adherent clumps or small ES cell-like colonies were formed in the primary culture [42, 46].

Culture of SSCs

Three components are considered essential for the maintenance of SSCs in culture: feeders, medium, and addition of growth factors.

In order to maintain cells in an undifferentiated state, mouse embryonic fibroblasts (MEF) or mouse testicular stromal cells (MTS) are used as feeder cells [43, 46].

The most commonly used culture media are the embryonic stem cell (ESC) medium or the StemPro® medium [44]. A serum-free condition has been described as being important to provide a good environment for the initial culture in vitro [47, 48], but the addition of fetal bovine serum (FBS) to the medium is used for long-term cultures.

Different exogenous agents are added to the culture medium to provide the culture with the appropriate signals in order to induce several cellular processes (Table 10.2).

Expansion of the SSCs Population In Vitro

The most critical period for the enrichment of SSCs is the first 2 weeks of culture. During this time, the formation of cell clumps loosely adhered to the plate is a good indication of GC morphology.

The first expansion should be mechanical or by gently pipetting up and down for few times to obtain a single cell suspension. The expansion ratio is then determined on the basis of the size and number of clumps. For subsequent expansions, cells are trypsinized and seeded into inactivated MEF or MTS.

The expansion rate for the second and third passages is about 1:1.5 or 1:2 (usually after 7 days). From three passages onward, the expansion rate ranges from 1:2 to 1:4 [49].

Transplantation of SSCs into In Vivo Systems

This step is essential for their functional characterization, but the stem cell potential depends on the recipient tissue. It is necessary to relocate these cells in the appropriate niche to maintain self-renewal and differentiation abilities [50].

Although different techniques exist, the basic procedure is the microinjection of the cell suspension into the seminiferous tubules of a receptive testis after GC depletion by busulfan treatment or chemotherapy. Under these conditions, SSCs are capable of colonizing the host testis by re-establishing the spermatogenesis process [39]. The efficacy of transplantation depends mainly on the number of SSCs injected into the testis and the quality of the recipient niche [51–53].

In the mouse model, most of the proliferating donor-derived cells are located on the basement membrane during the first post-transplant month. From the second month onward, GCs move from the basement membrane to the lumen and they differentiate into spermatozoa [54].

Therapy for Fertility Preservation

Currently, there are different established methods for fertility preservation in post-pubertal patients:

- Gonadal prevention: using shields during radiotherapy treatments with cancer patients to avoid directly striking the reproductive organs.
- Sperm banking: freezing healthy sperm before cancer treatment to use it in the future to conceive.
- Testis cancer surgery: nowadays, a usual procedure in surgical oncology is to remove only the cancerous nodule instead of the whole testis in selected patients with testis cancer.

- Testis tissue preservation: freezing testicular tissue from adult men is a valuable method to potentially preserve fertility since sperm may be found in the testis tissue despite the lack of sperm in the ejaculate.

In prepubertal boys, some of the above-mentioned strategies are unworkable because spermatozoa are lacking. In the human testis, spermatogenesis starts after puberty, so retrieval of spermatozoa and preservation of gonadal tissue before puberty are not feasible options to preserve fertility. However, there is a potential fertility preservation method for these patients:

- SSC banking: ejaculated sperm is not present in prepubertal boys with cancer. Nonetheless, it may be possible to freeze the SSC collected from the testis by means of a testicular biopsy before they receive cancer treatment, thus enabling fertility to be restored during their adult life after autologous transplantation:
- Young cancer patients who require chemotherapy and radiotherapy treatments present a blockade of the spermatogenesis process as a secondary effect.

The most helpful clinical strategy for retrieval of SSCs or GCs in such patients is to submit them to a testicular biopsy before starting the cancer treatment. This procedure consists in minor surgery to collect a piece of testicular tissue containing GCs and SSCs. These cells can be grown in vitro, and are then expanded and frozen in order to recover these patients' fertility in the future. Post-treatment would be the ideal time to transplant cells, as malignancy is absent, in order to provide them an appropriate niche for differentiation.

Conclusions

Isolation, maintenance and differentiation of SSCs form the next step in male fertility to restore spermatogenesis and to use SSCs in assisted reproduction. Since a small amount of tissue was collected by means of biopsy, the propagation of the SSCs in vitro to repopulate the testis becomes very important. The methodology developed and research conducted in human SSCs and GCs is very recent, and further research is needed.

References

1. Fathalla MF. Reproductive health: a global overview. Early Hum Dev. 1992;29:35–42.
2. Boivin J, Bunting L, Collins JA, Nygren KG. International estimates of infertility prevalence and treatment-seeking: potential need and demand for infertility medical care. Hum Reprod. 2007; 22:1506–12.
3. World Health Organization. WHO manual for the standardised investigation and diagnosis of the infertile couple. Cambridge: Cambridge University Press; 2000.
4. Thonneau P, Marchand S, Tallec A, Ferial ML, Ducot B, Lansac J, et al. Incidence and main causes of infertility in a resident population (1,850,000) of three French regions (1988–1989). Hum Reprod. 1991; 6:811–6.

5. Bahadur G, Ralph D. Gonadal tissue cryopreservation in boys with pediatric cancers. Hum Reprod. 1999;14:11–7.
6. Daugaard G, Hansen HH, Rurth M. Treatment of malignant germ cell tumors. Ann Oncol. 1990; 1: 195–202.
7. Pasqualotto FF, Rossi LM, Guilherme P, Ortiz V, Iaconelli Jr A, Borges Jr E. Etiology-specific outcomes of intracytoplasmic sperm injection in azoospermic patients. Fertil Steril. 2005;83(3):606–11.
8. Stephen EH, Chandra A. Declining estimates of infertility in the United States: 1982–2002. Fertil Steril. 2006;86:516–23.
9. Jarow JP, Espeland MA, Lipshultz LI. Evaluation of the azoospermic patient. J Urol. 1989;142:62–5.
10. Silber SJ, Nagy Z, Devroey P, Tournaye H, Van Seirteghem AC. Distribution of spermatogenesis in the testicles of azoospermic men: the presence or absence of spermatids in the testes of men with germinal failure. Hum Reprod. 1997;12:2422–8.
11. Silber SJ, Rodrigues-Rigau LJ. Quantitative analysis of testicle biopsy: determination of partial obstruction and prediction of sperm count after surgery for obstruction. Fertil Steril. 1981;36:480–5.
12. Turek PJ, Cha I, Ljung B-M. Systematic fine-needle aspiration of the testis: correlation to biopsy and results of organ "mapping" for mature sperm in azoospermic men. Urology. 1997;49:743–8.
13. Gorgy A, Podsiadly BT, Bates S, Craft IL. Testicular sperm aspiration (TESA): the appropriate technique. Hum Reprod. 1998;13:1111–3.
14. Choi J, Koh E, Suzuki H, Maeda Y, Yoshida A, Namiki M. Alu sequence variants of the BPY2 gene in proven fertile and infertile men with Sertoli cell-only phenotype. Int J Urol. 2007;14(5):431–5.
15. Mancini M, Carmignani L, Gazzano G, Sagone P, Gadda F, Bosari S, et al. High prevalence of testicular cancer in azoospermic men without spermatogenesis. Hum Reprod. 2007;22(4):1042–6.
16. Thomson JA, Itskovitz-Eldor J, Shapiro SS, Waknitz MA, Swiergiel JJ, Marshall VS, et al. Embryonic stem cell lines derived from human blastocysts. Science. 1998;282:1145–7.
17. Ilic D, Giritharan G, Zdravkovic T, Caceres E, Genbacev O, Fisher SJ, et al. Derivation of human embryonic stem cell lines from biopsied blastomeres on human feeders with minimal exposure to xenomaterials. Stem Cells Dev. 2009;18(9):1343–50.
18. Aguilar-Gallardo C, Poo M, Gomez E, Galan A, Sanchez E, Marques-Mari A, et al. Derivation, characterization, differentiation, and registration of seven human embryonic stem cell lines (VAL-3, -4, -5, -6 M, -7, -8, and −9) on human feeder. In Vitro Cell Dev Biol Anim. 2010;46(3–4):317–26.
19. Katz J, Keenan B, Snyder EY. Culture and manipulation of neural stem cells. Adv Exp Med Biol. 2010;671:13–22.
20. Goodman JW, Hodgson GS. Evidence for stem cells in the peripheral blood of mice. Blood. 1962;19:702–14.
21. Barnes DW, Loutit JF. Haemopoietic stem cells in the peripheral blood. Lancet. 1967;2(7526):1138–41.
22. Pittenger MF, Mackay AM, Beck SC, Jaiswal RK, Douglas R, Mosca JD, et al. Multilineage potential of adult human mesenchymal stem cells. Science. 1999;284(5411):143–7.
23. Zuk PA. The adipose-derived stem cell: looking back and looking ahead. Mol Biol Cell. 2010;21(11): 1783–7.
24. Toma JG, Akhavan M, Fernandes KJ, Barnabé-Heider F, Sadikot A, Kaplan DR, et al. Isolation of multipotent adult stem cells from the dermis of mammalian skin. Nat Cell Biol. 2001;3(9):778–84.
25. Alonso L, Fuchs E. Stem cells of the skin epithelium. Proc Natl Acad Sci USA. 2003;100 Suppl 1: 11830–5.
26. Guan K, Nayernia K, Maier LS, Wagner S, Dressel R, Lee JH, et al. Pluripotency of spermatogonial stem cells from adult mouse testis. Nature. 2006;440 (7088): 1199–203.

27. Fortier LA. Stem cells: classifications, controversies, and clinical applications. Vet Surg. 2005;34(5):415–23.
28. Olive V, Cuzin F. The spermatogonial stem cells: from basic knowledge to transgenic technology. Int J Biochem Cell Biol. 2005;37:246–50.
29. Nayernia K, Li M, Engel W. Spermatogonial stem cells. In: Schatten H, editor. Germ cell protocols: methods in molecular biology, vol. 253. Totowa: Humana Press; 2003. p. 105–20.
30. Brinster RL, Zimmermann JW. Spermatogenesis following male germ-cell transplantation. Proc Natl Acad Sci USA. 1994;9:11298–302.
31. Lawson KA, Hage WJ. Clonal analysis of the origin of primordial germ cells in the mouse. Ciba Found Symp. 1994;182:68–91.
32. McLaren A. Germ and somatic cell lineages in the developing gonad. Mol Cell Endocrinol. 2000;163: 3–9.
33. Donovan PJ, Stott D, Cairns LA, et al. Migratory and postmigratory mouse primordial germ cells behave differently in culture. Cell. 1986;44:831–8.
34. de Rooij DG, Grootegoed JA. Spermatogonial stem cells. Curr Opin Cell Biol. 1998;10: 694–701.
35. Nayernia K, Li M, Engel W. Spermatogial stem cells. Methods Mol Biol. 2004;253:105–20.
36. Russell LD, Ettlin RA, Hikim APS, Clegg ED. Histological and histopathological evaluation of the testis. Clearwater, FL: Cache River Press; 1990.
37. Tegelenbosch RA, de Rooij DG. A quantitative study of spermatogonial multiplication and stem cell renewal in the C3H/101 F1 hybrid mouse. Mutat Res. 1993;290:193–200.
38. Hofmann MC. Gdnf signaling pathways within the mammalian spermatogonial stem cell niche. Mol Cell Endocrinol. 2008;288:95–103.
39. Oatley JM, Brinster RL. Regulation of spermatogonial stem cell self-renewal in mammals. Annu Rev Cell Dev Biol. 2008;24:263–86.
40. Conrad S, Renninger M, Hennenlotter J, Wiesner T, Just L, Bonin M, et al. Generation of pluripotent stem cells from adult human testis. Nature. 2008;456(7220): 344–9.
41. Ko K, Araúzo-Bravo MJ, Tapia N, Kim J, Lin Q, Bernemann C, et al. Human adult germline stem cells in question. Nature. 2010;465(7301):E1. discussion E3.
42. Golestaneh N, Kokkinaki M, Pant D, Jiang J, DeStefano D, Fernandez-Bueno C, et al. Pluripotent stem cells derived from adult human testes. Stem Cells Dev. 2009;18(8):1115–26.
43. Kossack N, Meneses J, Shefi S, Nguyen HN, Chavez S, Nicholas C, et al. Isolation and characterization of pluripotent human spermatogonial stem cell-derived cells. Stem Cells. 2009;27(1):138–49.
44. He Z, Kokkinaki M, Jiang J, Dobrinski I, Dym M. Isolation, characterization, and culture of human spermatogonia. Biol Reprod. 2010;82(2):363–72.
45. Mizrak SC, Chikhovskaya JV, Sadri-Ardekani H, van Daalen S, Korver CM, Hovingh SE, et al. Embryonic stem cell-like cells derived from adult human testis. Hum Reprod. 2010; 25(1):158–67.
46. Guan K, Wolf F, Becker A, Engel W, Nayernia K, Hasenfuss G. Isolation and cultivation of stem cells from adult mouse testes. Nat Protoc. 2009;4(2): 143–54.
47. Kubota H, Avarbock MR, Brinster RL. Culture conditions and single growth factors affect fate determination of mouse spermatogonial stem cells. Biol Reprod. 2004;71:722–31.
48. Kubota H, Avarbock MR, Brinster RL. Growth factors essential for self-renewal and expansion of mouse spermatogonial stem cells. Proc Natl Acad Sci USA. 2004;101:16489–94.
49. Oatley JM, Brinster RL. Spermatogonial stem cells. Meth Enzymol. 2006;419:259–82.
50. Wyns C, Curaba M, Vanabelle B, Van Langendonckt A, Donnez J. Options for fertility preservation in prepubertal boys. Hum Reprod Update. 2010;16(3): 312–28.
51. Dobrinski I, Ogawa T, Avarbock MR, Brinster RL. Computer assisted image analysis to assess colonization of recipient seminiferous tubules by spermatogonial stem cells from transgenic donor mice. Mol Reprod Dev. 1999;53:142–8.
52. Ogawa T, Dobrinski I, Avarbock MR, Brinster RL. Transplantation of male germ line stem cells restores fertility in infertile mice. Nat Med. 2000;6:29–34.

53. Ohta H, Yomogida K, Dohmae K, Nishimune Y. Regulation of proliferation and differentiation in spermatogonial stem cells: the role of c-kit and its ligand SCF. Development. 2000;127: 2125–31.
54. Nagano M, Avarbock MR, Brinster RL. Pattern and kinetics of mouse donor spermatogonial stem cell colonization in recipient testes. Biol Reprod. 1999;60(6):1429–36.
55. Shinohara T, Avarbock MR, Brinster RL. Beta1- and alpha6-integrin are surface markers on mouse spermatogonial stem cells. Proc Natl Acad Sci USA. 1999;96(10):5504–9.
56. Kanatsu-Shinohara M, Toyokuni S, Shinohara T. CD9 is a surface marker on mouse and rat male germline stem cells. Biol Reprod. 2004;70(1):70–5.
57. Ryu BY, Orwig KE, Kubota H, Avarbock MR, Brinster RL. Phenotypic and functional characteristics of spermatogonial stem cells in rats. Dev Biol. 2004;274(1): 158–70.
58. Tolkunova EN, Malashicheva AB, Chikhirzhina EV, Kostyleva EI, Zeng W, Luo J, et al. E cadherin as a novel surface marker of spermatogonial stem cells. Cell Tissue Biol. 2009; 3(2):103–9.
59. Buageaw A, Sukhwani M, Ben-Yehudah A, Ehmcke J, Rawe VY, Pholpramool C, et al. GDNF family receptor alpha1 phenotype of spermatogonial stem cells in immature mouse testes. Biol Reprod. 2005;73(5):1011–6.
60. Ebata KT, Zhang X, Nagano MC. Expression patterns of cell-surface molecules on male germ line stem cells during postnatal mouse development. Mol Reprod Dev. 2005;72(2):171–81.
61. Seandel M, Falciatori I, Shmelkov SV, Kim J, James D, Rafii S. Niche players: spermatogonial progenitors marked by GPR125. Cell Cycle. 2008;7(2):135–40.
62. Seandel M, James D, Shmelkov SV, Falciatori I, Kim J, Chavala S, et al. Generation of functional multipotent adult stem cells from GPR125+ germline progenitors. Nature. 2007; 449(7160):346–50.
63. Tadokoro Y, Yomogida K, Ohta H, et al. Homeostatic regulation of germinal stem cell proliferation by the GDNF/FSH pathway. Mech Dev. 2002;113:29–39.
64. Hamra FK, Chapman KM, Nguyen DM, Williams-Stephens AA, Hammer RE, Garbers DL. Self renewal, expansion, and transfection of rat spermatogonial stem cells in culture. Proc Natl Acad Sci USA. 2005;102(48):17430–5.
65. Brinster RL. Germline stem cell transplantation and transgenesis. Science. 2002;296(5576):2174–6. Review.
66. Hofmann MC, Braydich-Stolle L, Dym M. Isolation of male germ-line stem cells; influence of GDNF. Dev Biol. 2005;279:114–24.
67. de Rooij DG. The spermatogonial stem cell niche. Microsc Res Tech. 2009;72:580–5.
68. Kanatsu-Shinohara M, Miki H, Inoue K, Ogonuki N, Toyokuni S, Ogura A, et al. Long-term culture of mouse male germline stem cells under serum-or feeder-free conditions. Biol Reprod. 2005;72(4): 985–91.
69. Kanatsu-Shinohara M, Inoue K, Ogonuki O, Miki H, Yoshida S, Toyokuni S, et al. Leukemia inhibitory factor enhances formation of germ cell colonies in neonatal mouse testis culture. Biol Reprod. 2007;76:55–62.
70. Pellegrini M, Grimaldi P, Rossi P, et al. Developmental expression of BMP4/ALK3/SMAD5 signaling pathway in the mouse testis: a potential role of BMP4 in spermatogonia differentiation. J Cell Sci. 2003;116: 3363–72.
71. Maki CB, Pacchiarotti J, Ramos T, Pascual M, Pham J, Kinjo J, et al. Phenotypic and molecular characterization of spermatogonial stem cells in adult primate testes. Hum Reprod. 2009;24(6):1480–91.
72. Kubota H, Avarbock MR, Brinster RL. Spermatogonial stem cells share some, but not all, phenotypic and functional characteristics with other stem cells. Proc Natl Acad Sci USA. 2003;100(11):6487–92.

Chapter 11
Approach to Fertility Preservation in Adult and Pre-pubertal Males

Fnu Deepinder and Ashok Agarwal

In the past two decades, major strides have been made in the curability of cancers leading to striking improvements in the chances of long-term survival, particularly in young men including children and adolescents. The use of chemotherapeutic and radio-therapeutic interventions has led to 70% survival rates in children with malignancies [1].

However, one of the major complications of these advanced treatment modalities is sterility and loss of gonadal function [2]. The effects might be transient or permanent depending upon the individual variability in the sensitivity to reproductive damage [3]. The severity of damage is dependent on the type of chemotherapy or radiotherapy, the treatment protocol, and the age of patients [4].

Although, future fertility of young males is very low on the relative quality of life parameters list at the time of anti-cancer treatment, infertility becomes an important issue following cure from cancer. According to a recent survey, 51% of men with cancer wanted children in future, including 77% of men who were childless when their cancer was first diagnosed [5]. Since it is difficult to predict who will survive or become infertile after anti-cancer treatment, fertility conservation is an important issue for those young cancer patients who have not yet started or completed their family size.

Although, fertility preservation in adult men by sperm cryopreservation is already established, options in pre-pubertal males are still experimental. There has been a tremendous progress in the development of strategies for germ cell banking. This chapter discusses the approach for fertility preservation in pre-pubertal and adult males including a number of technical and ethical concerns requiring careful evaluation.

A. Agarwal, Ph.D., HCLD (ABB) (✉)
Center for Reproductive Medicine, Glickman Urological and Kidney Institute,
OB-GYN and Women's Health Institute, Cleveland Clinic, Cleveland, OH, USA
e-mail: agarwaa@ccf.org

E. Seli and A. Agarwal (eds.), *Fertility Preservation in Males: Emerging Technologies and Clinical Applications*, DOI 10.1007/978-1-4614-5620-9_11, © Springer Science+Business Media New York 2012

Ethical and Legal Prospects

Informed Consent/Assent

The question of preserving fertility beyond a cancer patient's current treatment raises the need for an informed consent. This may be complicated because of the paucity of meaningful options and very little time available to most of the patients for taking any decision. Informed consent process for minor needs the involvement of patient's parents or legal guardians. Assent (permission less than full consent) is required in case of minors who are able to understand the issue, such as postpubertal boys and girls, together with the parental consent. However, for children too young to give an assent, parents may consent to experimental procedures only if the expected benefits are sufficient to justify the risks involved. In a recent study done in Netherlands, asking opinion of 318 parents of boys surviving cancer, sperm collection was approved by 70%; whereas spermatogonial stem cell collection by biopsy and hemicastration got only 61 and 33% approval, respectively [6]. The principles of beneficence and non-maleficence are considered paramount in such cases and oftentimes hospital ethics committee are asked to review parental decisions. Informed consent requires that the patient is given information that a reasonable person would want to know, and in enough detail that a reasonable person would be able to understand the procedure. Failure to offer all the existing methods of fertility preservation and accurately explaining the associated risks with such procedures may give rise to medical malpractice claims against health care providers. Hence it is essential to keep written documentation of such a process [7]. During assent, age appropriate information about sexual reproduction should be reviewed with patients as per their level of understanding, preferably by a psychologist or psychiatrist with expertise in children. In children enrolled in experimental fertility preservation protocols, the consent process should be performed in two stages. The decision to harvest immature germ cells made at the time of cancer treatment would rely on the parents or guardians. However, the decision of how to use the gametes could be made at a future time by the patient when he attains adulthood [8].

Risks Involved

A common cause of concern for both fertility specialists and cancer survivors seeking fertility preservation is whether their offspring are at higher risks for physical defects and cancer because of the effects of their disease, anti-cancer therapy, and cryopreservation techniques. Children born with disabilities may allege medical negligence in connection with their parent's fertility preservation during cancer treatment that preserved their life [9]. Men should be advised of a possible, not yet quantifiable, higher risk of genetic damage in sperm stored after diagnosis of cancer or initiation of cancer therapy. In non-cancer populations, there is no evidence of an increased risk of adverse outcomes if cryopreserved rather than fresh sperm are used for assisted reproduction. Pre-pubertal children may inadvertently get castrated because of loss of gonadal tissue during the collection of gametes in addition to the gonadotoxic chemotherapy and radiation. Hence children and their parents must be made aware of the risk of premature gonadal failure and delayed sexual development before proceeding with the fertility preservation options. Some experts have also questioned if it is ethical to enable cancer patients to reproduce as they face a greatly lowered life span, thus leaving a minor child bereft of one patient [10]. Furthermore, providers who store human genetic material for future use may face liability for damages in the event of loss or destruction of the cryopreserved tissue [9].

Religious Beliefs

Religious prohibitions against collection of genetic tissue for future use by natural or artificial methods may present another obstacle for fertility preservation. Beliefs and opinions vary across different religions and even among members of the same religion. In case of children, parental views should be respected, although a child's religious views may be different or may change over the time [11].

Financial Costs

Patients also need to be made aware of the financial costs involved as the insurance companies do not always cover the costs of cryopreservation.

Legal Parenthood

Challenges may arise with respect to legal parentage of the children resulting from cryopreserved tissues. The need to determine legal paternity arises in context of inheritance and federal benefits. Most courts recognize that if a man dies before placement of his gametes; he is not a parent of the child resulting from posthumous assisted reproduction, unless he consented otherwise. Hence it is important to include in the informed consent the patient's intended legal relationship to any resulting child.

Future Disposition in Case of Death

Discussing fertility preservation with young males and their families takes patience and sensitivity. In addition to addressing patient's own future use of his gamete, it also involves the consent for future disposition of that tissue or its use by a specific partner or parent in case of patient's death. Because of the important ethical and emotional issues raised, it is advisable to have a bioethicist available to talk to patients along with their families, and formulate institutional policies to provide limits and guidance [12]. In the absence of prior written directives, it is recommended that ownership of the tissue should not be transferred to the patient's relatives in case the patient dies prior to use of the stored germ cell material. However, many women have successfully gained access to their deceased husbands' sperms convincing the courts that the storage of gametes by adults itself implied a decision to procreate. Furthermore, complicated issues may also arise in situations such as when the patient dies without banking the semen and the surviving partner or parent requests the health care provider for posthumous extraction of sperm. As compared to adults, a child's rights get violated if his gametes are used posthumously as he had a limited understanding of the process at the time of collection of his gametes. The two-step consent process ensures that the stored genetic material cannot be used for reproductive purposes until the child becomes an adult and is able to give full consent [9].

Semen Collection

Although post-pubertal males are ordinarily capable of ejaculation and provide sperm for storage, teenagers may be embarrassed to discuss the option of masturbation in front of their parents. A mental health expert, oncology nurse, or a social worker can minimize the embarrassment by discussing it outside the presence of their parents. With parental permission, having some non-violent erotic magazines or videos in the collection room may be helpful. For boys who cannot ejaculate, invasive procedures can be done with their assent and parental consent [12]. If a young teen objects to any of the above procedures, they should not be done, despite parental wishes [13].

Established Options for Fertility Preservation

Modification of Treatment Regime

The first line of fertility protection is to reduce the exposure of gonads to cytotoxic chemotherapy and radiotherapy. Wherever possible, radiation exposure to the gonads should be limited by using radiation shields. Gonadal damage can also be reduced by reduction in the dose and frequency of cytotoxic chemotherapy, or by substitution with less gonadotoxic agents.

In a recent study of 355 adult patients treated for Hodgkin's disease, impairment of spermatogenesis was seen in 8% of patients receiving nonalkylating chemotherapy in contrast to 60% of patients who received alkylating agents. Moreover, recovery of spermatogenesis occurred in 82% of patients treated with non-alkylating chemotherapy as compared to just 30% in the alkylating group [14]. Another trial in Hodgkin's disease patients compared gonadal toxicity of combination chemotherapy with alkylating agents containing MOPP regime (mechlormethine, procarbazine, vincristine, and prednisone) vs. non-alkylating chemotherapy ABVD (doxorubicin, bleomycin, vinblastine, and dacarbazine). In this trial, 86% of the patients treated with MOPP demonstrated persistent azoospermia as compared to ABVD regime, in which all patients showed recovery of spermatogenesis [15]. This substantiates that cancer treatment can potentially be modified in some circumstances to reduce gonadal toxicity.

Sperm Banking

Traditionally, sperm banking by cryopreservation of at least three semen samples with an abstinence period of at least 48 h in between the samples has been recommended for adult males desiring to preserve their fertility [16]. However, many young cancer patients, especially those with testicular carcinoma or Hodgkin's disease already have decreased semen quality at the time of diagnosis and start of anti-cancer therapy [17]. Freezing and thawing semen further reduces sperm's count, motility, and viability [18]. Additional samples and longer abstinence periods may be used to achieve higher total sperm counts. Moreover, recent advances in assisted reproductive technology especially the advent of intracytoplasmic sperm injection (ICSI) have made it possible for a man to become a father even if only a few spermatozoa remain alive after cryopreservation [19, 20].

Semen banking should ideally be done before the start of cancer treatment. Theoretically semen collection and storage is feasible after the initiation of chemotherapy and radiation therapy, at least until azoospermia ensues. However, it is advisable to wait for 12–18 months because of the time taken for the recovery of spermatogenesis and significant increase in the frequency of sperm aneuploidy persisting for 18 months or more after initiation of anti-cancer treatment [21–23].

The semen collection process is achieved by masturbation. The patient should be provided a sterile specimen collection cup and ample time and privacy to produce sample. It is important to avoid lubricants such as petroleum jelly and saliva as these substances may be spermatotoxic. If no ejaculate is expelled on climax, then a post-ejaculate urinalysis should be done to assess for retrograde ejaculation. If retrograde ejaculation is observed, alpha agonist medications may be administered to convert retrograde to antegrade ejaculation. If this is not successful, then alkalization of the urine and subsequent collection and processing of the postejaculate urine sample may facilitate isolation of viable sperm. If the patient is unable to reach climax, care should be taken to ensure that he has had ample privacy and time [24].

Although post-pubertal males are generally able to ejaculate, some young cancer patients may not be able to produce a sample by masturbation. A strong vibrator or a rectal electric probe can be used to stimulate ejaculation in these boys; however it

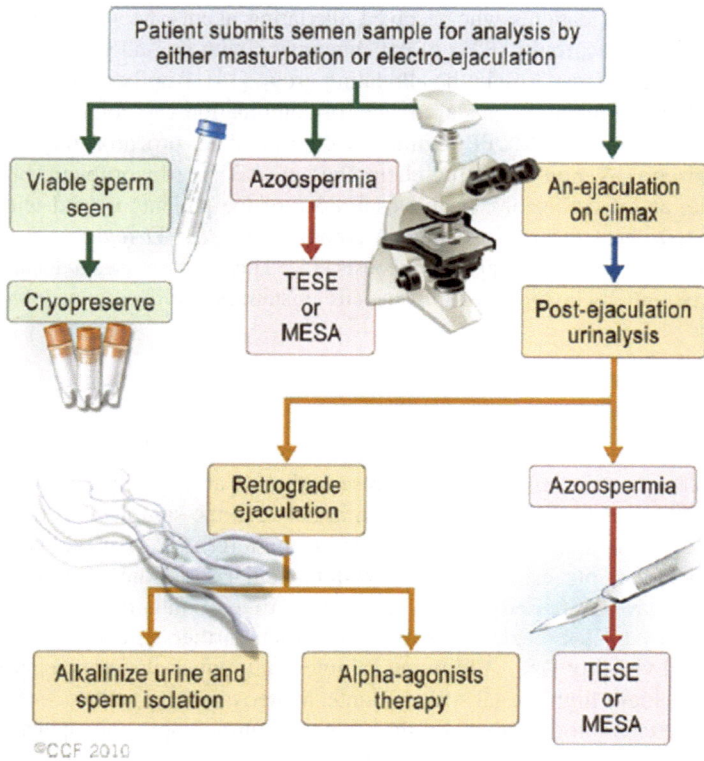

Fig. 11.1 Approach for sperm banking in post-pubertal and adult males

should be used under anesthesia to avoid pain [25]. Advanced methods for sperm retrieval include microsurgical epididymal sperm aspiration (MESA) for patients with un-reconstructable obstructive azoospermia [26] and testicular sperm extraction (TESE) for patients suffering from nonobstructive azoospermia [27–29]. Although TESE in combination with ICSI has shown some recent promise in patients with post-chemotherapy azoospermia, long-term potential for genetic risks among children born with this technique are still unknown (Fig. 11.1).

Experimental Options for Fertility Preservation

Absence of spermatozoa and spermatids in testis prevents pre-pubertal boys to benefit from the technique of sperm freezing. Spermarche, the production of sperm, occurs at approximately 13–14 years of age. Although few spermatids might be present in some seminiferous tubules in the late pre-pubertal period; such spermatocytes never seem to give rise to spermatozoa [30]. Furthermore, cryopreserved semen samples are a finite source and do not offer the patients a chance to achieve natural fertility. Hence, with the recent advances in assisted reproductive techniques, mature and immature sperm extraction and maturation [31, 32], focus has been shifted to the possibility of testicular tissue and germ cell preservation in young cancer patients.

Spermatogonial Stem Cell Transplantation

The testicular tissue contains testicular precursors called spermatogonial stem cells. These cells are the genuine totipotent population of cells in the adult body and they undergo self-renewal throughout the life. These are also more resistant than other testicular cells to variety of toxic insults [33]. The spermatogonial stem cells are present in pre-pubertal testicular tissue, and can be isolated and successfully cryo-preserved with almost 70% cells surviving freezing and thawing as demonstrated in animal experiments [34]. These stem cells can be re-transplanted autologously into the testis, where they recolonize the seminiferous tubules, generating complete spermatogenesis and mature germ cells thus restoring natural fertility (spermatogonial stem cell autotransplantation) [35–38]. Animal studies done so far have suggested that there is flexibility with respect to the donor age because there are only modest differences in stem cell concentration and colony expansion among newborn, pre-adolescent, and adult testes. However, recipient age has a significant impact on donor germ cell engraftment with the preadolescent testes demonstrating significantly higher spermatogenesis as compared to adults [38, 39]. Therefore, spermatogonial stem cell transplantation is a potentially viable method for fertility preservation in both pre-pubertal and adult males.

Although, successful establishment of fertility has been achieved with stem cell transplantation in mice, rats, and goats, the technique is still experimental in humans [35–38]. One of the crucial steps for re-fertilization is the safe retrieval of sufficient testicular tissue before the cytotoxic insults of chemotherapy and radiotherapy. A safe cryopreservation protocol for the spermatogenic stem cells, similar to freezing of human spermatozoa and embryos is also needed. Dimethylsulfoxide (DMSO), as a cryoprotectant has been shown to maintain the structure of testicular tissue especially the spermatogonia better than others like propanediol and glycerol. In a study involving 16 infertile men, after freezing with 5% DMSO, $70\pm6\%$ of seminiferous tubules were found to be good as compared to $37\pm3\%$ with propanediol [40]. However, DMSO itself is a potential carcinogen which restricts its application in clinical practice [41, 42]. Although DMSO is usually used in concentrations of 10–20% [32, 43, 44], Keros et al. have shown promising results with 5% DMSO in human testicular tissue cryopreservation [40, 45]. Another potential concern is the contamination of spermatogonia with tumor cells leading to retransmission of cancer back to the recipient. Systemic malignancies can metastasize to the testis, and transplantation of testicular cells can re-expose the recipients to the same problem. Presence of malignant cells in the gonadal tissue can be screened by various cytogenetic and molecular markers. However, if there is clinical evidence of testicular tissue involvement, or when the potential for occult metastasis is high, auto-transplantation of germ cells is not recommended. Transplantation of human spermatogonial stem cells into animals may be performed instead (spermatogonial stem cell xeno-transplantation) [46, 47]. However, animal infectious agents such as retroviruses may be introduced in human germ line when these cells are used to procure conception [48]. Furthermore, it is impossible to generate haploid male gametes from diploid germ cells with the existing in vitro approaches as the testicular stem cells expansion and meiotic entry appears to be blocked in cultures of testicular cell suspension [49]. Some of the other challenges encountered with this technology include the ischemic damage to the transplanted testicular tissue, in vitro enrichment of stem cell spermatogonia, and non-invasive transfer of germ cell suspensions into the rete testis. Hence, in spite of latest developments in the field suggesting bright prospects for fertility preservation in young cancer patients who have not achieved puberty, techniques still need to be developed for isolation, storage, and re-infusion of spermatogonial stem cells in humans.

Testicular Tissue Cryopreservation

This technique involves removal of testicular tissue from the patient before cytotoxic therapy and transplantation of testicular tissue pieces to an ectopic site such as under the skin of patients after successful completion of cancer therapy (ectopic auto-grafting of testicular tissue) or into animals (ectopic xeno-grafting of testicular tissue). The grafted testicular tissue revascularizes in the ectopic site producing complete spermatogenesis, first demonstrated by Brinster and Zimmermann in mice [50, 51]. Sperm retrieval from grafted tissue can be used to generate healthy offspring with assisted fertilization techniques. Grafting has been successfully demonstrated in hamsters, pigs, goats, calves, rabbits, and monkeys [52–54]. In a recently published on-going trial, Children's hospital of Philadelphia has been successful in enrolling 16 pre-pubertal boys aged 3 months to 14 years for testicular cryopreservation as a potential future use in an experimental protocol. In this study, 76% (16/21) of the parents of pre-pubertal boys with cancer consented for testicular biopsy. Of the 16 patients enrolled, 14 underwent the procedure without any negative intra- or post-operative sequelae [55]. A Belgian study has earlier demonstrated survival and proliferative capacity of spermatogonia and sertoli cells in cryopreserved immature human cryptorchid tissue transplanted into mice scrotum through orthotopic xenografting model [56]. Lately, Sato et al. from Japan demonstrated successful spermatogenesis in the xenograft of human infant testicular tissue grafted into nude mice until pachytene spermatocyte stage [57]. In addition to spermatogenic, the steroidogenic function of the testicular tissue has also been shown to restore with testicular grafting [53]. However, grafting of adult testicular tissue has shown only limited success in contrast to immature tissue. Furthermore, autografting of cryopreserved testicular tissue share similar concerns of cancer recurrence as spermatogonial stem cell auto-transplantation. Likewise, as previously discussed, xenografting of human gonadal tissue poses significant risks of introducing various animal infectious agents in human germ line.

Hormonal Gonado-Protection

This strategy of fertility preservation in cancer patients involves manipulation of the hypothalamic pituitary gonadal axis (HPG axis) to render the testes less susceptible to cytotoxic therapy. An initial hypothesis was that suppression of the HPG axis prior to cytotoxic therapy using agents such as sex steroids, gonadotropin releasing hormone (GnRH) agonists or antagonists would protect the gonad from cytotoxic damage by preventing the germ cells from actively proliferating thus inducing resting or a less sensitive state. Lately, it has been shown that GnRH agonists release the germ cells from the block on differentiation that would otherwise occur [58]. Furthermore, it has also been proposed that GnRH antagonists reduce the high levels of intra-testicular testosterone caused by radiation therapy, leading to a reduction in interstitial volume and testicular edema and thus allowing early resumption of spermatogenesis [59].

In rats, spermatogenesis and fertility has shown to be restored following treatment with radiation or chemotherapy by suppressing testosterone with GnRH agonists or antagonists either before or after the cytotoxic insult [60, 61]. However, hormonal therapy has not been successful in preserving fertility or speed recovery of spermatogenesis in men and other primates [62–64]. The only successful human study evaluated testosterone alone in men without cancer treated with cyclophosphamide for nephrotic disorders [65]. Thus despite the success in restoring fertility by hormone treatment of rats rendered azoospermic by chemotherapy and radiation, the application of this procedure to humans is uncertain [66].

Selection of Appropriate Technique for Adult and Pre-pubertal Males

Studies have revealed that most young male patients with cancer do not seek fertility preservation [67, 68]. Reasons for this apparent underutilization are either lack of timely information and referral by physicians, or psychological, logistic and financial constraints of the patients [5, 12]. Even when sperm is banked, less than 30% of men return to use their stored specimens [69, 70]. Preserving the fertility of young men requires coordinated efforts and attention to a host of ethical and legal issues by oncologists and fertility specialists. Physicians have a responsibility to inform patients that infertility is a potential risk of their cancer treatment. Basic questions such as whether fertility preservation options decrease their chance of successful cancer treatment, increase the risk of maternal or perinatal complications, or compromise the health of offspring should be answered. Informed consent should be obtained from the patients regarding the type of fertility preservation treatment and legal status of any resulting child. The more specifically and comprehensively a written directive is executed by the patient during his lifetime, the less likely it will be that disputes will arise regarding future use of the stored genetic material.

Table 11.1 Summary for fertility preservation options in pre-pubertal and adult males

	Sperm cryopreservation	Spermatogonial stem cell autotransplantation	Spermatogonial stem cell xenotransplantation	Testicular tissue cryopreservation and ectopic autografting	Testicular tissue cryopreservation and ectopic xenografting
Eligible candidates	Adult male	Pre-pubertal male	Pre-pubertal male	Pre-pubertal male	Pre-pubertal male
Method of pregnancy	Assisted reproduction	Natural intercourse or assisted reproduction	Assisted reproduction	Assisted reproduction	Assisted reproduction
Critical comments	Unfeasible in prepubertal males, finite source, cannot achieve natural fertility	Invasive procedure, risk of retransmission of cancer	Invasive procedure, risk of transfer of animal infections into human germ line	Invasive procedure, risk of retransmission of cancer	Invasive procedure, risk of transfer of animal infections into human germ line
Status of procedure	Established	Investigational	Investigational	Investigational	Investigational

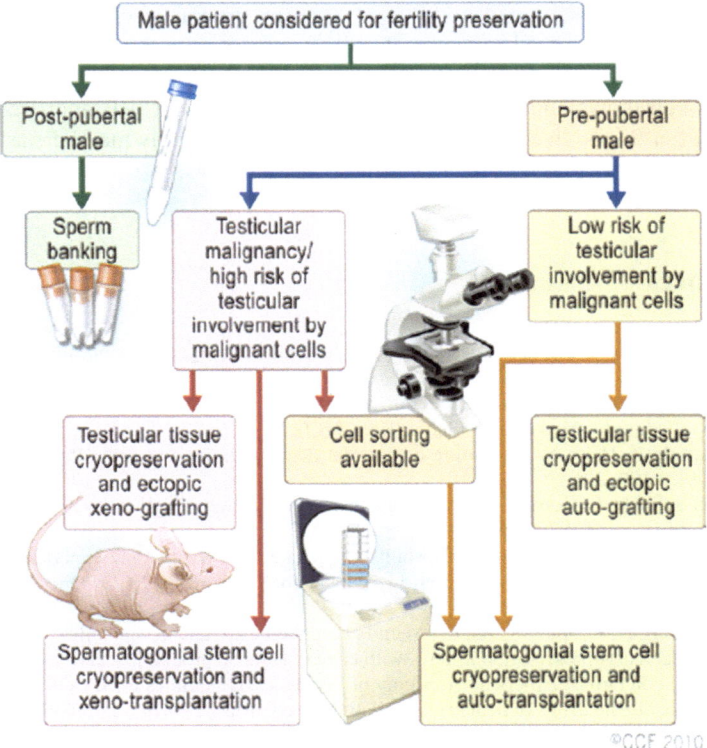

Fig. 11.2 Selection of appropriate established or experimental technique for fertility preservation in adult and pre-pubertal males

Literature suggests that sperm banking is the only effective method of fertility preservation in men. Cryopreservation of sperm before initiation of cytotoxic therapy is the best method of preserving fertility in post-pubertal males. Testicular or spermatogonial stem cell cryopreservation and auto-transplantation or xenografting are still investigational. However, these are the only options available for pre-pubertal boys. Gonadal protection through hormone manipulation is ineffective [49, 71, 72]. Table 11.1 summarizes available fertility preservation options for pre- and postpubertal males. Figure 11.2 provides a flowchart for selection of appropriate fertility preservation technique among the available established and experimental options.

Conclusions

Recent advances in medicine have led to ever increasing number of pre-pubertal and adult males surviving cancer treatment. This has increased the need to improve the existing technology for cryopreservation of gametes and search for new fertility preservation options. As of today, only sperm cryopreservation is considered accepted standard clinical practices. Fertility preservation options in pre-pubertal males are still experimental. Cryopreservation of testicular tissue and spermatogonial stem cell transplantation should only be offered within IRB-approved clinical protocol after thorough counseling of patients and their family members as there are still many unresolved issues related to these technologies.

References

1. Robison LL. Methodologic issues in the study of second malignant neoplasms and pregnancy outcomes. Med Pediatr Oncol Suppl. 1996;1:41–4.
2. Mackie EJ, Radford M, Shalet SM. Gonadal function following chemotherapy for childhood Hodgkin's disease. Med Pediatr Oncol. 1996;27(2):74–8.
3. Blumenfeld Z, Haim N. Prevention of gonadal damage during cytotoxic therapy. Ann Med. 1997;29(3):199–206.
4. Howell S, Shalet S. Gonadal damage from chemotherapy and radiotherapy. Endocrinol Metab Clin North Am. 1998;27(4):927–43.
5. Schover LR, Brey K, Lichtin A, Lipshultz LI, Jeha S. Knowledge and experience regarding cancer, infertility, and sperm banking in younger male survivors. J Clin Oncol. 2002;20(7): 1880–9.
6. van den Berg H, Repping S, van der Veen F. Parental desire and acceptability of spermatogonial stem cell cryopreservation in boys with cancer. Hum Reprod. 2007;22(2):594–7.
7. Committee on Drugs, American Academy of Pediatrics. Guidelines for the ethical conduct of studies to evaluate drugs in pediatric populations. Pediatrics. 1995;95(2):286–94.
8. Bahadur G. Ethics of testicular stem cell medicine. Hum Reprod. 2004;19(12):2702–10.
9. Crockin SL. Legal issues related to parenthood after cancer. J Natl Cancer Inst Monogr. 2005;34:111–3.
10. Robertson JA. Procreative liberty and harm to offspring in assisted reproduction. Am J Law Med. 2004;30(1):7–40.
11. Stegmann BJ. Unique ethical and legal implications of fertility preservation research in the pediatric population. Fertil Steril. 2010;93(4):1037–9.
12. Schover LR, Agarwal A, Thomas Jr AJ. Cryopreservation of gametes in young patients with cancer. J Pediatr Hematol Oncol. 1998;20(5):426–8.
13. Robertson JA. Cancer and fertility: ethical and legal challenges. J Natl Cancer Inst Monogr. 2005;34:104–6.
14. van der Kaaij MA, Heutte N, Le Stang N, et al. Gonadal function in males after chemotherapy for early-stage Hodgkin's lymphoma treated in four subsequent trials by the European Organisation for Research and Treatment of Cancer: EORTC Lymphoma Group and the Groupe d'Etude des Lymphomes de l'Adulte. J Clin Oncol. 2007;25(19):2825–32.
15. Viviani S, Santoro A, Ragni G, Bonfante V, Bestetti O, Bonadonna G. Gonadal toxicity after combination chemotherapy for Hodgkin's disease. Comparative results of MOPP vs ABVD. Eur J Cancer Clin Oncol. 1985;21(5):601–5.
16. Meseguer M, Molina N, Garcia-Velasco JA, Remohi J, Pellicer A, Garrido N. Sperm cryopreservation in oncological patients: a 14-year follow-up study. Fertil Steril. 2006;85(3):640–5.
17. Agarwal A, Shekarriz M, Sidhu RK, Thomas Jr AJ. Value of clinical diagnosis in predicting the quality of cryopreserved sperm from cancer patients. J Urol. 1996;155(3):934–8.
18. Gandini L, Lombardo F, Lenzi A, Spano M, Dondero F. Cryopreservation and sperm DNA integrity. Cell Tissue Bank. 2006;7(2):91–8.
19. Hallak J, Hendin BN, Thomas Jr AJ, Agarwal A. Investigation of fertilizing capacity of cryopreserved spermatozoa from patients with cancer. J Urol. 1998;159(4):1217–20.

20. Kuczynski W, Dhont M, Grygoruk C, Grochowski D, Wolczynski S, Szamatowicz M. The outcome of intracytoplasmic injection of fresh and cryopreserved ejaculated spermatozoa – a prospective randomized study. Hum Reprod. 2001;16(10):2109–13.
21. De Mas P, Daudin M, Vincent MC, et al. Increased aneuploidy in spermatozoa from testicular tumour patients after chemotherapy with cisplatin, etoposide and bleomycin. Hum Reprod. 2001;16(6):1204–8.
22. Howell SJ, Shalet SM. Spermatogenesis after cancer treatment: damage and recovery. J Natl Cancer Inst Monogr. 2005;34:12–7.
23. Shin D, Lo KC, Lipshultz LI. Treatment options for the infertile male with cancer. J Natl Cancer Inst Monogr. 2005;34:48–50.
24. Brannigan RE. Fertility preservation in adult male cancer patients. Cancer Treat Res. 2007;138:28–49.
25. Ohl DA, Wolf LJ, Menge AC, et al. Electroejaculation and assisted reproductive technologies in the treatment of anejaculatory infertility. Fertil Steril. 2001;76(6):1249–55.
26. Janzen N, Goldstein M, Schlegel PN, Palermo GD, Rosenwaks Z, Hariprashad J. Use of electively cryopreserved microsurgically aspirated epididymal sperm with IVF and intracytoplasmic sperm injection for obstructive azoospermia. Fertil Steril. 2000;74(4):696–701.
27. Devroey P, Liu J, Nagy Z, et al. Pregnancies after testicular sperm extraction and intracytoplasmic sperm injection in non-obstructive azoospermia. Hum Reprod. 1995;10(6):1457–60.
28. Chan PT, Palermo GD, Veeck LL, Rosenwaks Z, Schlegel PN. Testicular sperm extraction combined with intracytoplasmic sperm injection in the treatment of men with persistent azoospermia postchemotherapy. Cancer. 2001;92(6):1632–7.
29. Palermo G, Joris H, Devroey P, Van Steirteghem AC. Pregnancies after intracytoplasmic injection of single spermatozoon into an oocyte. Lancet. 1992;340(8810):17–8.
30. Paniagua R, Nistal M. Morphological and histometric study of human spermatogonia from birth to the onset of puberty. J Anat. 1984;139(Pt 3):535–52.
31. Fishel S, Green S, Bishop M, et al. Pregnancy after intracytoplasmic injection of spermatid. Lancet. 1995;345(8965):1641–2.
32. Hovatta O, Foudila T, Siegberg R, Johansson K, von Smitten K, Reima I. Pregnancy resulting from intracytoplasmic injection of spermatozoa from a frozen thawed testicular biopsy specimen. Hum Reprod. 1996;11(11):2472–3.
33. Russell LD, Ettlin RA, Hakim AP. Mammalian spermatogenesis. Clearwater: Cache River Press; 1999.
34. Izadyar F, Matthijs-Rijsenbilt JJ, den Ouden K, Creemers LB, Woelders H, de Rooij DG. Development of a cryopreservation protocol for type A spermatogonia. J Androl. 2002;23(4):537–45.
35. Brinster RL, Avarbock MR. Germline transmission of donor haplotype following spermatogonial transplantation. Proc Natl Acad Sci U S A. 1994;91(24):11303–7.
36. Ogawa T, Dobrinski I, Avarbock MR, Brinster RL. Transplantation of male germ line stem cells restores fertility in infertile mice. Nat Med. 2000;6(1):29–34.
37. Honaramooz A, Behboodi E, Megee SO, et al. Fertility and germline transmission of donor haplotype following germ cell transplantation in immunocompetent goats. Biol Reprod. 2003;69(4):1260–4.
38. Ryu BY, Orwig KE, Avarbock MR, Brinster RL. Stem cell and niche development in the postnatal rat testis. Dev Biol. 2003;263(2):253–63.
39. Shinohara T, Orwig KE, Avarbock MR, Brinster RL. Remodeling of the postnatal mouse testis is accompanied by dramatic changes in stem cell number and niche accessibility. Proc Natl Acad Sci U S A. 2001;98(11):6186–91.
40. Keros V, Rosenlund B, Hultenby K, Aghajanova L, Levkov L, Hovatta O. Optimizing cryopreservation of human testicular tissue: comparison of protocols with glycerol, propanediol and dimethylsulphoxide as cryoprotectants. Hum Reprod. 2005;20(6):1676–87.
41. Avarbock MR, Brinster CJ, Brinster RL. Reconstitution of spermatogenesis from frozen spermatogonial stem cells. Nat Med. 1996;2(6):693–6.

42. Aslam I, Fishel S, Moore H, Dowell K, Thornton S. Fertility preservation of boys undergoing anti-cancer therapy: a review of the existing situation and prospects for the future. Hum Reprod. 2000;15(10):2154–9.
43. Shinohara T, Inoue K, Ogonuki N, et al. Birth of offspring following transplantation of cryopreserved immature testicular pieces and in-vitro microinsemination. Hum Reprod. 2002;17(12):3039–45.
44. Jezek D, Schulze W, Kalanj-Bognar S, Vukelic Z, Milavec-Puretic V, Krhen I. Effects of various cryopreservation media and freezing-thawing on the morphology of rat testicular biopsies. Andrologia. 2001;33(6):368–78.
45. Keros V, Hultenby K, Borgstrom B, Fridstrom M, Jahnukainen K, Hovatta O. Methods of cryopreservation of testicular tissue with viable spermatogonia in pre-pubertal boys undergoing gonadotoxic cancer treatment. Hum Reprod. 2007;22(5):1384–95.
46. Nagano M, Patrizio P, Brinster RL. Long-term survival of human spermatogonial stem cells in mouse testes. Fertil Steril. 2002;78(6):1225–33.
47. Sofikitis N. Transplantation of human spermatogonia into the seminiferous tubules (STs) of animal testicles results in the completion of the human meiosis and the generation of human motile spermatozoa. Fertil Steril. 1999;72(suppl 1):S83–4.
48. Patience C, Takeuchi Y, Weiss RA. Infection of human cells by an endogenous retrovirus of pigs. Nat Med. 1997;3(3):282–6.
49. Orwig KE, Schlatt S. Cryopreservation and transplantation of spermatogonia and testicular tissue for preservation of male fertility. J Natl Cancer Inst Monogr. 2005;34:51–6.
50. Brinster RL, Zimmermann JW. Spermatogenesis following male germ-cell transplantation. Proc Natl Acad Sci U S A. 1994;91(24):11298–302.
51. Ogawa T. Spermatogonial transplantation technique in spermatogenesis research. Int J Androl. 2000;23(Suppl 2):57–9.
52. Honaramooz A, Snedaker A, Boiani M, Scholer H, Dobrinski I, Schlatt S. Sperm from neonatal mammalian testes grafted in mice. Nature. 2002;418(6899):778–81.
53. Schlatt S, Kim SS, Gosden R. Spermatogenesis and steroidogenesis in mouse, hamster and monkey testicular tissue after cryopreservation and heterotopic grafting to castrated hosts. Reproduction. 2002;124(3):339–46.
54. Oatley JM, de Avila DM, Reeves JJ, McLean DJ. Spermatogenesis and germ cell transgene expression in xenografted bovine testicular tissue. Biol Reprod. 2004;71(2):494–501.
55. Ginsberg JP, Carlson CA, Lin K, et al. An experimental protocol for fertility preservation in prepubertal boys recently diagnosed with cancer: a report of acceptability and safety. Hum Reprod. 2010;25(1):37–41.
56. Wyns C, Curaba M, Martinez-Madrid B, Van Langendonckt A, Francois-Xavier W, Donnez J. Spermatogonial survival after cryopreservation and short-term orthotopic immature human cryptorchid testicular tissue grafting to immunodeficient mice. Hum Reprod. 2007;22(6):1603–11.
57. Sato Y, Nozawa S, Yoshiike M, Arai M, Sasaki C, Iwamoto T. Xenografting of testicular tissue from an infant human donor results in accelerated testicular maturation. Hum Reprod. 2010;25(5):1113–22.
58. Meistrich ML, Wilson G, Kangasniemi M, Huhtaniemi I. Mechanism of protection of rat spermatogenesis by hormonal pretreatment: stimulation of spermatogonial differentiation after irradiation. J Androl. 2000;21(3):464–9.
59. Porter KL, Shetty G, Meistrich ML. Testicular edema is associated with spermatogonial arrest in irradiated rats. Endocrinology. 2006;147(3):1297–305.
60. Meistrich ML, Kangasniemi M. Hormone treatment after irradiation stimulates recovery of rat spermatogenesis from surviving spermatogonia. J Androl. 1997;18(1):80–7.
61. Meistrich ML, Wilson G, Huhtaniemi I. Hormonal treatment after cytotoxic therapy stimulates recovery of spermatogenesis. Cancer Res. 1999;59(15):3557–60.
62. Johnson DH, Linde R, Hainsworth JD, et al. Effect of a luteinizing hormone releasing hormone agonist given during combination chemotherapy on posttherapy fertility in male patients with lymphoma: preliminary observations. Blood. 1985;65(4):832–6.

63. Waxman JH, Ahmed R, Smith D, et al. Failure to preserve fertility in patients with Hodgkin's disease. Cancer Chemother Pharmacol. 1987;19(2):159–62.
64. Thomson AB, Anderson RA, Irvine DS, Kelnar CJ, Sharpe RM, Wallace WH. Investigation of suppression of the hypothalamic-pituitary-gonadal axis to restore spermatogenesis in azoospermic men treated for childhood cancer. Hum Reprod. 2002;17(7):1715–23.
65. Masala A, Faedda R, Alagna S, et al. Use of testosterone to prevent cyclophosphamide-induced azoospermia. Ann Intern Med. 1997;126(4):292–5.
66. Shetty G, Meistrich ML. Hormonal approaches to preservation and restoration of male fertility after cancer treatment. J Natl Cancer Inst Monogr. 2005;34:36–9.
67. Wallace WH, Anderson RA, Irvine DS. Fertility preservation for young patients with cancer: who is at risk and what can be offered? Lancet Oncol. 2005;6(4):209–18.
68. Chung K, Irani J, Knee G. Sperm cryopreservation for male patients with cancer: an epidemiological analysis at the University of Pennsylvania. Eur J Obstet Gynecol Reprod Biol. 2004;113(Suppl 1):S7–11.
69. Audrins P, Holden CA, McLachlan RI, Kovacs GT. Semen storage for special purposes at Monash IVF from 1977 to 1997. Fertil Steril. 1999;72(1):179–81.
70. Blackhall FH, Atkinson AD, Maaya MB, et al. Semen cryopreservation, utilisation and reproductive outcome in men treated for Hodgkin's disease. Br J Cancer. 2002;87(4):381–4.
71. Lee SJ, Schover LR, Partridge AH, et al. American Society of Clinical Oncology recommendations on fertility preservation in cancer patients. J Clin Oncol. 2006;24(18):2917–31.
72. Jeruss JS, Woodruff TK. Preservation of fertility in patients with cancer. N Engl J Med. 2009;360(9):902–11.

Protocols

Chapter 12
Sperm Preparation and Freezing for Banking

Pankaj Talwar

Semen Freezing Protocol

Here, we describe the *liquid nitrogen vapor technique* for cryofreezing the spermatozoa. The technique is simple, easy to learn, and does not require expensive equipment.

Sample Collection

The semen collection is achieved via masturbation in a healthy man.

A. The patient should be counseled and provided with a sterile specimen collection jar and ample time and privacy to do so.
B. Avoidance of lubricants (soap, jelly, and saliva) is important, as these are spermatotoxic.
C. If no ejaculate is achieved by the patient, then a postejaculate urinalysis should be examined to review for retrograde ejaculation.
D. If retrograde ejaculation is observed, alpha agonists may be administered in an effort to convert retrograde to antegrade ejaculation. Alkalinization of the urine may be carried out and urine post ejaculation may be used to harvest the sperms.
E. If the patient is unable to reach climax or have errection, audio-video aids may be used.
F. If this difficulty persists, then consideration should be given to vibratory stimulation, electroejaculation.
G. Surgical testicular sperm extraction techniques may have a potential role in such patients.

P. Talwar (✉)
ART Centre, Army Hospital Research and Referral, Dhaula Kuan,
New Delhi 110010, India
e-mail: pankaj_1310@yahoo.co.in

E. Seli and A. Agarwal (eds.), *Fertility Preservation in Males: Emerging Technologies and Clinical Applications*, DOI 10.1007/978-1-4614-5620-9_12,
© Springer Science+Business Media New York 2012

Liquid Nitrogen Vapor Cooling

Cryovials

The sample is collected in the designated sample collection room of the center to avoid contamination and temperature-related changes in the semen sample.

A. Confirm the particulars of the patient and indication of the cryofreezing on receiving the sample from the patient. Keep the semen sample at room temp for 30 min for liquefaction. Label the cryovials and make the entries in the logbook.
B. After liquefaction, measure the total volume of the ejaculate and carry out semen analysis as per WHO guidelines. If the sample is satisfactory, then it can be frozen raw. Concentrating the sperm prior to freezing can enhance postthaw recovery of semen samples with low sperm counts. Samples in cases of surgical sperm retrieval techniques or those with poor counts are prepared by double-density gradient or other suitable methods and frozen. Sperm pellet may also be frozen if the swim up does not have adequate number of sperms or sperms have poor morphological scoring.
C. Take out 2–3 mL of semen freezing media from the bottle at 4–8°C under laminar flow hood and keep in the incubator at room temperature.
D. Ensure that both semen sample and sperm freezing medium (SFM) are at room temperature. Dilute the semen 1:1 (v/v) with the SFM. The medium should be added dropwise to the semen and the solution carefully mixed after each additional drop of SFM. This is done as glycerol is toxic for the sperms. The procedure should be carried out under laminar flow hood. The mixture is left at room temperature for a period of 10–15 min to equilibrate.
E. Load the diluted semen into straws or cryovials and seal according to the manufacturer's recommendations (Figs. 12.1 and 12.2). It is very important to leave some air space in the lower part of the straw for sealing as well as to allow expansion of the solution during freezing.
F. Suspend the straws horizontally for 30 min, just above the surface of the liquid nitrogen. Cryovials should be attached to a cane and then suspended above the surface of the liquid nitrogen for the same period of time.
G. Finally, transfer the straws or cryovials into liquid nitrogen and store at −196°C (Fig. 12.3).

For Thawing the Sample

A. Remove straws or cryovials from liquid nitrogen and wash the vial under running water for few minutes. Now place them at room temperature till the time the sweating gets over and the sample has liquefied completely.

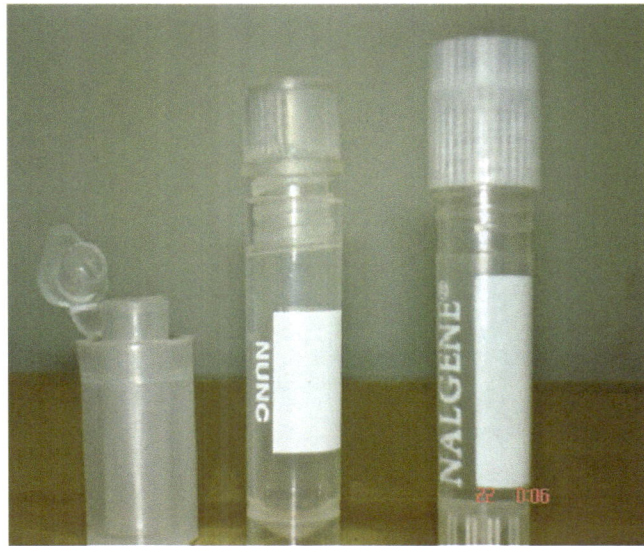

Fig. 12.1 Commercially available vials for semen packaging. Cryotubes may have internal or external threads. Few of these may have an external reservoir for liquid nitrogen for better temperature maintenance

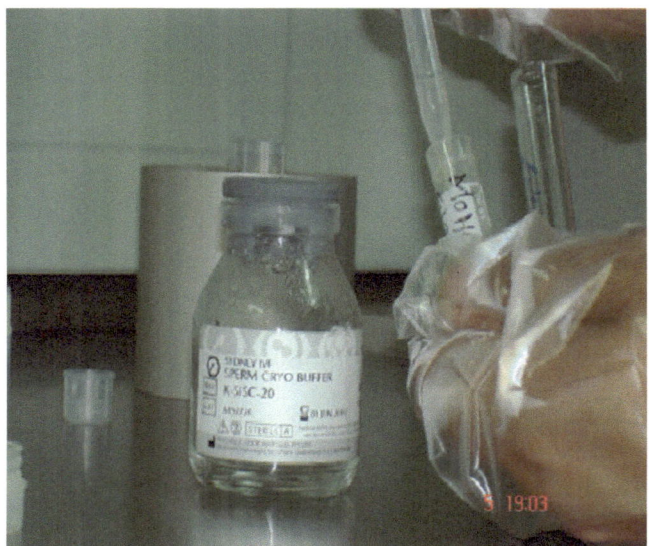

Fig. 12.2 Glycerolated semen being loaded in the cryovials

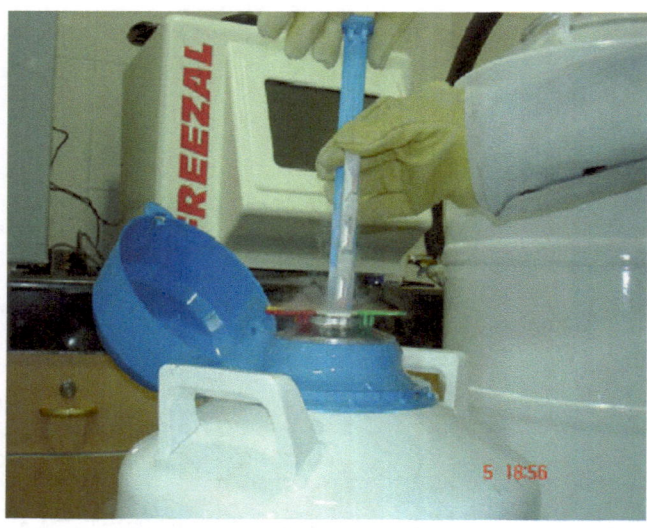

Fig. 12.3 Semen containing vials being loaded on the aluminum canes before being dipped in LN$_2$ in the cryocontainers

B. Wipe the vials totally dry and open the cap of the cryovials according to the manufacturer's instructions and remove the thawed semen.
C. Dilute the semen with sperm preparation medium (1:1) to reduce the toxic effect of glycerol.
D. Quickly evaluate the survival of the sperm. If necessary, thaw additional cryovials/straws.
E. Place the specimen in incubator for 15 min.
F. Immediately prepare sample by the density gradient method, swim up, or single wash technique with sperm preparation medium.
G. A final concentration of minimum 10×10^6/mL is recommended.

Cryostraws

A. Washed and prepared semen sample is filled in the prelabeled 0.5-mL, clear, flexible, ionomeric resin straws (CBS) using aspirating device (Figs. 12.4 and 12.5).
B. Straws are sealed using SYMS sealer (Fig. 12.6).
C. Sealed straws are suspended 4–5 cm above liquid nitrogen level in a Styrofoam box horizontally (Fig. 12.7)
D. The straw should not be touching each other and they should be minimum 0.2–0.3 cm apart for proper circulation of the liquid nitrogen vapor around the straws.

Fig. 12.4 Daisy goblets, visotubes, and straws for semen packaging. Color coding helps in easy identification of the sample

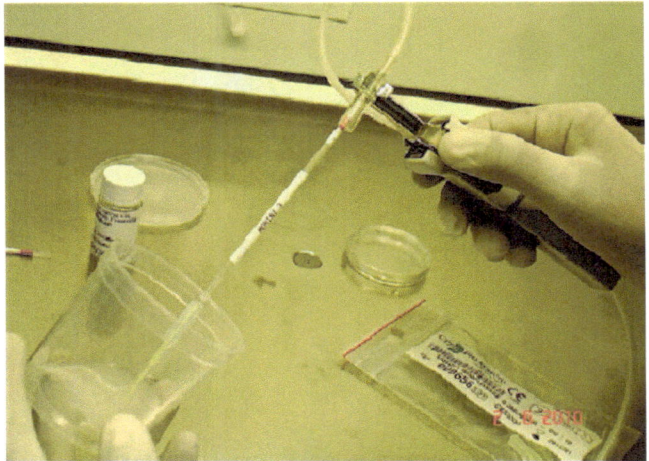

Fig. 12.5 Raw semen sample being loaded in the labeled CBS cryostraws using a manual aspirator

E. Straws are cooled for 10–20 min. By this time, the contents of the straws have frozen.
F. Now plunge them into liquid nitrogen and transferred in the predecided goblets in LN_2 container and taken on inventory (Fig. 12.8).

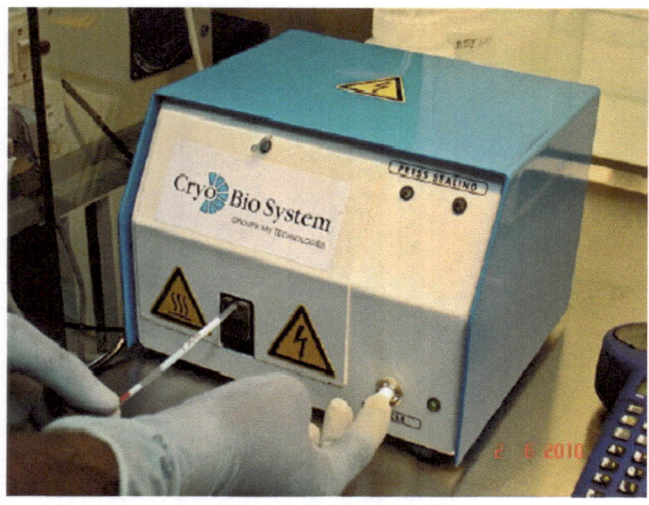

Fig. 12.6 SYMS CBS: Straw sealer being used to seal the labeled and loaded straws

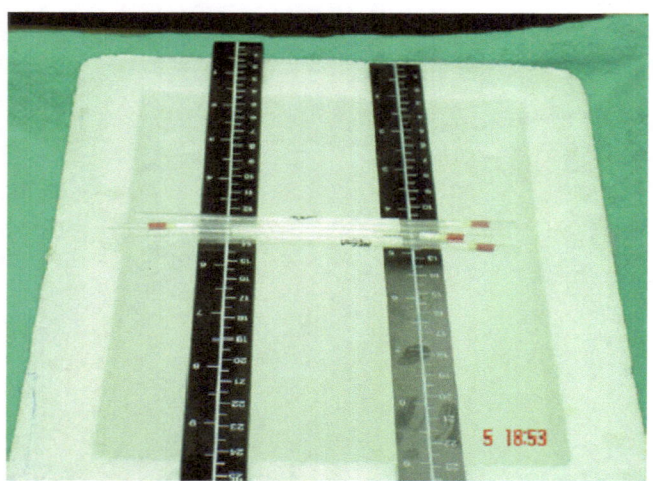

Fig. 12.7 Vapor-phase cooling of the semen loaded straws being done using a Styrofoam box containing liquid nitrogen

Technique Using Controlled Rate Freezing Method

Computer-controlled freezers. (a) Liquid nitrogen vapor-filled chambers (e.g., Kryo-10, Planer, Sunbury, UK and Nicool Models, Air Liquide, Bussy-Saint-Georges, France); (b) cooled metal blocks – CryoLogic, Australia – are widely used cryoplanes.

A. Ensure that both the sample and sperm cryopreservation buffer are at room temperature.

Fig. 12.8 Storage
of goblets containing
semen-loaded straws
in liquid nitrogen tank
with temperature
and level alarms

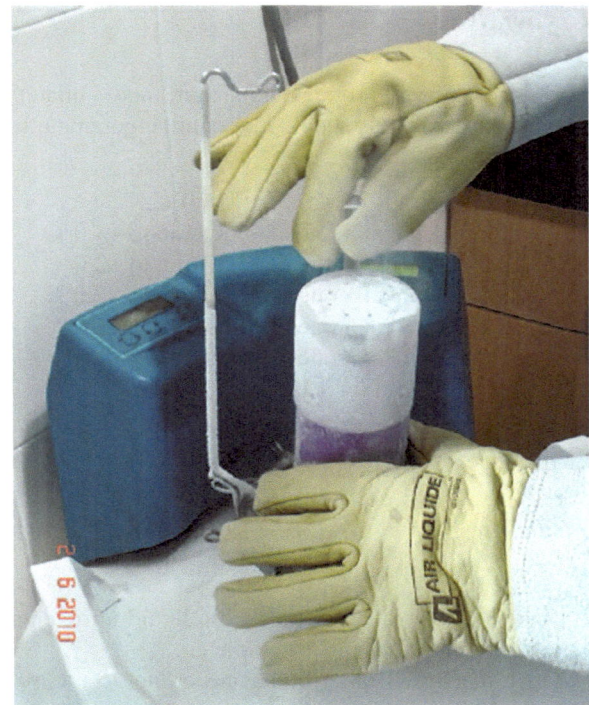

B. Mix two volumes of sperm cryopreservation buffer to one volume of sample.
C. Leave mixture for 10 min at room temperature.
D. Label straws with relevant information.
E. Load the sample into a freezing straw or cryovial and seal according to manufacturer's instructions.

Straws

Load straws into freezing machine and initiate freeze program. The program for straws should have similar parameters to that given below:

- Start temperature is 20°C.
- Cooling rate of 6°C/min until −80°C.
- At −80°C, plunge the straws into the liquid nitrogen.

Cryovials

Load cryovials into freezing machine and initiate freeze program. The freeze program for cryovials should have similar parameters to those given below:

- Start temperature is 20°C.
- Cooling rate of −0.5°C/min to +5.0°C.
- At +5.0°C, cool at a rate of −1°C/min to +4.0°C.
- At +4.0°C, cool at a rate of −2°C/min to +3.0°C.
- At +3.0°C, cool at a rate of −4°C/min to +2.0°C.
- At +2.0°C, cool at a rate of −8°C/min to +1.0°C.
- At +1.0°C, cool at a rate of −10°C/min to −80.0°C.
- At −80.0°C, hold for 10 min.
- Plunge into liquid nitrogen.

Thawing

A. Remove straws or cryovials from liquid nitrogen and place them at room temperature until thawing is complete.
B. Open the straws or cryotubes according to the manufacturer's instructions and remove the thawed semen.
C. Dilute the semen with HEPES buffer (1:1) to reduce the toxic effect of glycerol.
D. Quickly evaluate the survival of the sperm. If necessary, thaw additional straws for preparation.
E. Immediately prepare the thawed sample by the density gradient/single wash and swim up method.

Chapter 13
Preservation of Sperm Isolates and Testicular Biopsy Samples for Banking

Bhushan K. Gangrade

Protocol for Isolation and Cryopreservation of Testicular Sperm

Materials and Equipment

Sterile culture dishes
Polystyrene conical tubes
Sterile glass pipettes
Syringes (50 mL)
Syringe (3 mL) with 21 gauze needle
Syringe filter (Nalgene, 0.22 μm)
Microscope glass slides
Cover glass
Gloves
Protective goggles
Sterile pair of scissors
Sterile pair of forceps
Centrifuge
Microscope
Refrigerator
Weighing chemical balance
Nunc Cryo tube vials
Aluminum canes
Plastic cryosleeves
Liquid nitrogen storage dewar
Liquid nitrogen

B.K. Gangrade, Ph.D. (✉)
IVF Laboratory, Center for Reproductive Medicine,
3435 Pinehurst Avenue, Orlando, FL 32804, USA
e-mail: bkgangrade@hotmail.com

E. Seli and A. Agarwal (eds.), *Fertility Preservation in Males: Emerging Technologies and Clinical Applications*, DOI 10.1007/978-1-4614-5620-9_13,
© Springer Science+Business Media New York 2012

Reagents

Sperm Washing Medium (modified HTF with human serum albumin, 5 mg/mL; SAGE IVF Inc., Trumbull, CT, USA).

Sperm Freezing Medium (TEST yolk buffer with gentamycin sulfate; Irvine Scientific, Santa Ana, CA, USA).

Tissue Culture Grade Water (SAGE IVF Inc., Trumbull, CT, USA).

Erythrocyte (RBC) LYSIS BUFFEr (155 mM ammonium chloride, 10 mM potassium bicarbonate, 2 mM EDTA; pH 7.2).

Mineral oil.

Preparation of Erythrocyte (RBC) Lysis Buffer

Weigh the following chemicals separately.

NH_4Q (Sigma Cat #A 0171)	0.829 g
$KHCO_3$ (Sigma Cat #P 9144)	0.100 g
EDTA (Sigma Cat #ED2SS)	0.074 g

Dissolve in 100 mL of tissue culture grade water and adjust the pH to 7.2. Filter sterilize using Nalgene syringe filters (0.22 µm) attached to 50-mL syringe. The RBC lysis buffer may be stored in refrigerator (4°C) for 1 month.

Specimen

Note: Patient identification and labeling of the specimen container should be performed as per standard laboratory protocol. Sterile techniques and universal precautions should be exercised while processing the tissue sample.

The testicular biopsy sample is surgically retrieved by the surgeon or the urologist. The tissue is collected in a sterile petri dish in 2.0 mL of sperm washing medium. If the surgery is performed at a distant facility and the tissue needs to be transported to the Andrology Laboratory for processing, the testicular tissue may be transferred to a sterile container with tight lid. At all times, the tissue should be completely submerged in the sperm washing medium. Care should be taken to avoid exposure to extreme temperature during transport. The testicular tissue can be kept at room temperature for short (1–2 h) duration without any detrimental effect on sperm parameters.

Procedure

A. Wash the testicular tissue with 1–2 mL of sperm washing medium (hereafter referred as "culture medium") to remove red blood cells.

B. Transfer the tissue to a petri dish and add a few drops of culture medium to keep the tissue moist. Using a pair of sterile scissor, mince the tissue.

C. Add 1.0 mL of culture medium to the finely minced sample. Aspirate the suspension and gently pass it through a 21 gauze needle attached to a 3.0-mL syringe. Repeat this process 2–3 times. The passage through the needle breaks down the seminiferous tubules in small pieces and also dislodges the sperm from the lumen.

D. Place 5–10 μL of suspension in petri dish and overlay with mineral oil. Observe and record the number of spermatozoa per high power field (200×).

E. Transfer the suspension to a conical tube and let it stand at room temperature for 5 min. This allows the pieces of seminiferous tubules and large tissue clumps to settle down.

F. Transfer the supernatant to another conical tube and centrifuge (300× g) for 10 min.

G. Discard the supernatant.

H. If the pellet exhibits the presence of red blood cells (as it invariably does), add 2.0 mL of RBC lysis buffer to the tube and resuspend the pellet.

I. Centrifuge at 400× g for 5 min.

J. Discard the supernatant and wash the pellet with culture medium (1.0 mL).

K. Resuspend the pellet in 0.2-mL culture medium.

L. Add equal volume (0.2 mL) of TEST-Yolk buffer (freezing medium) slowly to the sperm suspension.

M. Transfer the above mixture (sperm suspension and freezing medium) to a prelabeled Nunc cryovial.

N. Place the cryovial in a refrigerator (4°C) for 45 min.

O. Expose the cryovial to the liquid nitrogen vapor phase (20–30 cm above the liquid nitrogen surface) for 45 min. This can be achieved by placing the vial in a wire mesh container hanging in the neck of a Dewar tank.

P. Place the cryovial on the aluminum cane and immerse in liquid nitrogen Dewar for storage.

Q. Record the location in the Dewar inventory.

Chapter 14
Testis Tissue Xenografting

Jose R. Rodriguez-Sosa, Stefan Schlatt, and Ina Dobrinski

Testis tissue xenografting involves a series of procedures from donor tissue preparation to graft recovery for analysis and/or sperm harvesting. In the current section, we describe these protocols in detail (modified from Dobrinski and Rathi [1]).

Materials

Preparation of Donor Tissue

1. Phosphate-buffered saline (PBS).
2. Dulbecco's modified Eagle's medium (DMEM) (or other balanced cell culture medium).
3. Plastic tissue culture dishes (60×15 and 100×20 mm).
4. Dissection instruments: Iris forceps (~10 cm, 0.8-mm tips), small forceps (~11 cm, 0.1×0.6 mm tips), curved iris forceps (~10 cm, 0.8 mm tips, curved), dissecting scissors (~10 cm, straight), disposable scalpel (#10).

Cyropreservation of Donor Tissue

1. Freezing medium: Any suitable cyoprotectant medium and method can be used. Here, for its practical use and the results obtained previously, we describe the use

I. Dobrinski, DVM, M.V.Sc., Ph.D. (✉)
Department of Comparative Biology and Experimental Medicine,
University of Calgary, 3300 Hospital Drive, Calgary, AB T2N 4N1, Canada
e-mail: idobrins@ucalgary.ca

E. Seli and A. Agarwal (eds.), *Fertility Preservation in Males: Emerging Technologies and Clinical Applications*, DOI 10.1007/978-1-4614-5620-9_14, © Springer Science+Business Media New York 2012

of DMSO-based medium and a conventional slow freezing method. Reagents: Heat-inactivated fetal calf serum (FCS), DMSO, and DMEM or other cell culture medium.
2. 2-mL cryovials.
3. Nalgene freezing container designed to provide a controlled cooling rate of 1 °C/min when placed into a −70 °C freezer.
4. 15-mL centrifuge tubes.
5. Plastic tissue culture dishes (60×15 and 100×20 mm).

Recipient Preparation and Ectopic Xenografting of Donor Tissue

1. Recipient mice: Immunodeficient (e.g., NCR nu/nu, SCID, or RAG) male mice, 6–8-weeks old. Female mice can also be used, but castration is easier in males.
2. Anesthesia reagents: Any suitable anesthetic can be used in accordance with animal care and use guidelines.
3. Tuberculin syringes (1 mL).
4. Injection needles (26½ gauge).
5. 70% ethanol.
6. Betadine solution.
7. Instruments suitable for mouse surgery.
8. Suture material with needle (6–0 Silk braid w/needle).
9. Wound clips (7.5 mm, e.g., Michel® clips).
10. Clip applying-removing forceps (12.7 cm, e.g., Michel®).
11. Heating pad.

Collection of Testis Xenografts for Analysis and Sperm Harvesting

1. Bouin's solution (or any other fixative, depending on the analysis method).
2. Sample vials.
3. Plastic tissue culture dishes (60×15 and 100×20 mm).
4. A pair of small forceps (~11 cm, 0.1×0.6 mm tips).
5. One pair of dissecting scissors (~10 cm, straight).
6. Culture medium or PBS.
7. Disposable scalpel (#10).
8. 40-μm cell strainer.
9. 15-mL centrifuge tubes. 10.
10. 70% ethanol.

Methods

Collection of Donor Tissue

1. Obtain testis tissue by castration or biopsy from a donor male.
2. Place testis in PBS or biopsies into culture medium, maintaining sterile conditions.
3. Keep the collected tissue on ice and transport to the laboratory.

Preparation of Donor Tissue

Prepare testis tissue in the tissue culture hood.

A. Wash each testis in ice-cold PBS containing antibiotics two to three times before transferring into a culture dish with PBS. In the case of biopsies, wash testis fragments two to three times with ice-cold culture medium containing antibiotics by spinning them down at 150 g for 2 min and resuspending in fresh ice- cold culture medium.
B. For intact testes, remove tunica vaginalis by making an incision along surface and extrude the testis. Proceed, then, to remove from testis all annex structures (spermatic cord, epididymis, connective tissue). Wash testes once in cold PBS and transfer them into a culture dish with PBS.

1. Carefully remove the tunica albuginea of the testis by using a scalpel blade and a pair of scissors. If the testis is very small, the tunica can be removed by squeezing the testicular tissue out of the tunica through a small incision made on one end while holding the tunica with a pair of small forceps on the other end.
2. Depending on the size of the testis, either the whole testis tissue can be cut into small pieces of around 1–2 mm^3 in size using curved forceps and a scalpel blade or large pieces of testis tissue can first be removed from the testis and then cut into smaller pieces. All this should be done in ice-cold culture medium and under sterile conditions in a small culture dish (60×15 mm).
3. Transfer the prepared tissue fragments to ice-cold culture medium in small culture dishes on ice until grafting.

C. For testis biopsies, these should be cut into 1–2 mm3 pieces and transferred to ice-cold culture medium in small culture dishes on ice until grafting.

Cryopreservation of Donor Tissue

Freezing

1. Prepare freezing medium by mixing FCS, DMEM, and DMSO at a ratio of 1:3:1.
2. Transfer approximately ten pieces of testis tissue into a 2-mL cryovial and add 0.5 mL of freezing medium at room temperature.

3. Place the vials into the Nalgene freezing container set at room temperature and introduce the container to a −70 °C freezer. Tissue fragments are left overnight into the container and freezer before being transferred into liquid nitrogen.

Thawing

1. Remove cryovials from liquid nitrogen and hold them at room temperature for 1 min to evaporate any remaining liquid nitrogen.
2. Introduce cryovials in a 25 °C water bath for 1 min.
3. Add ~1.5 mL into each vial, transfer content into a sterile 15-mL centrifuge tube, and 10 mL of fresh medium.
4. Centrifuge at 300 g for 2 min, discard supernatant, and resuspend tissue fragments in 10 mL of fresh medium.
5. Repeat wash to completely eliminate cryoprotectant.
6. Resuspend fragments in ~2 mL of fresh medium and transfer content to a tissue culture dish containing enough culture medium to maintain the tissue fragments in suspension. Maintain dish with fragments on ice until transplantation.

Recipient Preparation and Ectopic Xenografting of Donor Tissue

Anesthesia of Recipient Mouse

1. Weigh the mouse and anesthetize. Monitor anesthetic depth by the absence of voluntary and reflex movement.
2. Position the mouse in dorsal recumbency for castration (described below) on warm pad and keep mouse warm during the entire procedure.

Castration of Recipient Mouse

1. Prepare sterile surgical field by clipping or plucking the hair (not necessary in nude mice), wiping with 70% ethanol and betadine solution.
2. Make a 0.5–1-cm ventral midline skin incision to expose the abdominal wall.
3. Lift abdominal wall by using a small forceps at the point of the white line to avoid accidentally injuring abdominal organs and proceed to make a ~0.5-cm incision at the midline of the abdominal wall to expose the peritoneal cavity.
4. Use one iris forceps to hold the abdominal wall, and use another pair of iris forceps to search for the fat pads attached to the epididymis and testis in the peritoneal cavity. Gently pull the fat pad out until the testis is exteriorized and the testicular artery and epididymis are clearly visible.

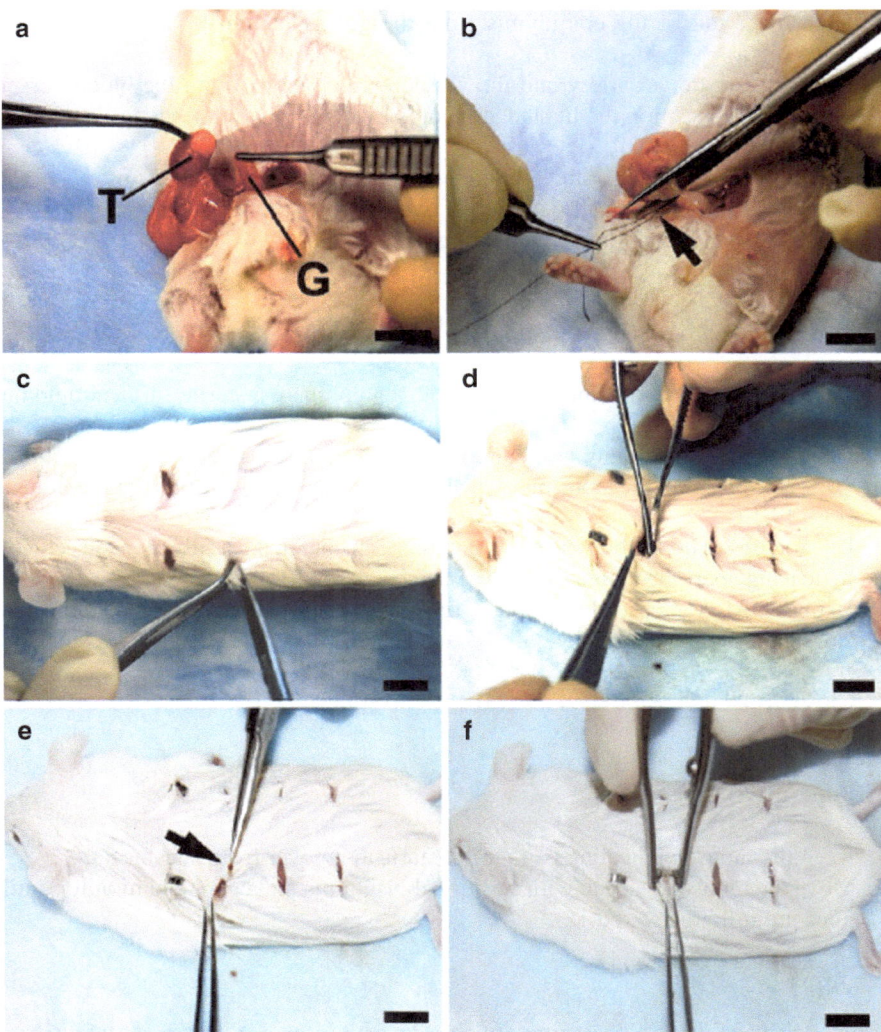

Fig. 14.1 Key methodological steps in recipient preparation and ectopic transplantation of testicular tissue. (**a**) Photograph illustrating the detachment of the testis (*T*) from the gubernaculum (*G*) prior to ligation of the testicular artery and annexed structures. (**b**) Once the testis has been detached from the gubernaculum, the testicular artery and vas deferens are ligated. In this photograph, the ligature has been already placed (*arrow*) and the sectioning of the ligated structures is about to be performed. (**c**) Once the mouse has been castrated and its back has been aseptically prepared, 0.5–1-cm incisions are made in the skin to introduce each testis fragment. (**d**) In order to place each testis fragment under the dorsal skin, the subcutaneous tissue is teased apart with small scissors to produce a small cavity. (**e**) The testis fragment is placed with fine forceps deep into the subcutaneous cavity by holding the border of the skin incision with small forceps to expose the cavity. (**f**) Once the testis fragment has been placed, the incision is closed with a Michel clip. Bars = 1 cm

5. Detach the tail of the epididymis from the gubernaculum by blunt dissection (Fig. 14.1a).
6. Ligate the testicular artery and the vas deferens together with the blood vessel with silk, and section the ligated structures by cutting between the testis and the ligature (Fig. 14.1b).
7. Repeat the procedure for the second testis.
8. Suture the abdominal wall with one or two surgical stitches.
9. Close the skin incision with one or two Michel clips.

Ectopic Xenografting

1. Position the mouse in ventral recumbency and prepare a sterile surgical field on its back as above.
2. Depending on how many grafts are to be inserted (generally 4–8/mouse), make ~0.5-cm-long skin incisions on each side of the midline of the back of the mouse (Fig. 14.1c).
3. Use forceps to hold a border of the skin incision, and make a subcutaneous cavity by teasing apart the connective tissue using scissors (Fig. 14.1d).
4. Using an iris forceps, place a piece of testis tissue deep into the subcutaneous cavity, holding the border of the skin incision with another iris forceps (Fig. 14.1e).
5. Close the skin incision with one Michel clip (Fig. 14.1f).

Postoperative Care

1. Keep the mouse on heating pad until it starts to recover from anesthesia.
2. Move the mouse to a cage with additional insulation and cover and monitor until mice are fully recovered.

Collection of Testis Xenografts for Analysis and Sperm Harvesting

Xenografts Recovery

1. Sacrifice the host mouse according to animal care and use guidelines.
2. Make a midline skin incision on the back skin running from the tail to the neck and open skin. This exposes the grafts which can be located either on the subcutaneous tissue or attached to the skin.
3. Carefully remove the grafts using a pair of forceps and a pair of scissors.
4. Record the number of grafts recovered and size and weight of individual grafts.
5. Retrieve the seminal vesicles from the abdomen of the mouse and record their weight as an indication of testosterone production by the grafted tissue.

For Histological Analysis

1. Suspend xenografts into sample vial containing Bouin's solution (or other fixative) in a volume ~10 timed that of the xenograft, and label vial appropriately.
2. Incubate overnight in the refrigerator followed by washing at least three times in 70% ethanol at intervals of 24 h preferably.
3. Proceed for processing and embedding in paraffin.

For Sperm Harvesting

1. Wash xenografts by spinning them down at 300 g for 1 min and resuspending them in culture medium containing antibiotics.
2. Cut grafts into small pieces and mince carefully with the forceps in a tissue culture dish containing 3–5 mL of culture medium.
3. Filter minced tissue through the 40-μm cell strainer.

Reference

1. Dobrinski I, Rathi R. Ectopic grafting of mammalian testis tissue into mouse hosts. In: Hou SX, Singh SR, editors. Methods in molecular biology germ line stem cells. Totowa: Humana Press; 2008. p. 139–48.

Index